Thoughts of a Scientist, Citizen, and Grandpa on Climate Change

Bridging the Gap between Scientific and Public Opinion

———————— O ————————

ERIC P. GRIMSRUD

iUniverse, Inc.
Bloomington

Thoughts of a Scientist, Citizen, and Grandpa on Climate Change
Bridging the Gap between Scientific and Public Opinion

Copyright © 2012 Eric P. Grimsrud

iUniverse Star
an iUniverse, Inc. imprint

iUniverse books may be ordered through booksellers or by contacting:

iUniverse
1663 Liberty Drive
Bloomington, IN 47403
www.iuniverse.com
1-800-Authors (1-800-288-4677)

ISBN: 978-1-938908-02-6 (sc)
ISBN: 978-1-938908-03-3 (e)

Library of Congress Control Number: 2012904513

Printed in the United States of America

iUniverse rev. date: 5/3/2012

For Charlie, Kate, Elsa, Emma, and Krista

The author wishes to thank his wife, Kathy, and his neighbor, Jerry Elwood, for their encouragement, feedback and editing during the preparation of this book.

Kathy, an avid reader and excellent wordsmith, has served as her husband's "first filter" on everything he has written for the public domain in recent years. As Kathy does not have a scientific background, any section she did not understand was considered not yet ready for public consumption and sent back to the drawing board.

A most fortunate coincidence during the author's recent existence in rural Montana is that Jerry Elwood, the former Director of Climate Change Research at the U.S. Department of Energy, also retired from his full time job and just happened to settle in the same neighborhood. Upon meeting Jerry the author immediately realized that he had a first-rate source of information and experience concerning climate change research literally next door and then did, indeed, take full advantage of Jerry's generous support of this project.

Contents

Introduction

As suggested by the title of this book, I have spent my professional life as a scientist; during much of it, I have done research in the specific area of atmospheric chemistry. As a result, I have acquired a reasonably good understanding of the various environmental issues that are directly related to Earth's atmosphere. I have also learned a lot about the scientific communities of our country and about how they come to their collective views. As a citizen of the U.S. during this same period — and perhaps because I was raised in a small-town newspaper family — I have also been very interested in the processes by which the general public comes to its collective views, some of which have concerned the same environmental issues I have studied. As also suggested by this book's title, my wife and I are now grandparents, due to the arrivals of Charlie, Kate, Elsa, Emma, and Krista during the last five years. Therefore, we are, of course, intensely interested in the well-being of these youngsters, as well as that of their future siblings, cousins, friends, and descendants. All three of these past and present interests have merged to provide the motivation for this book.

The subject of global warming has been of great interest to both the scientific community and the general public during the last two decades and is certain to become of central importance to our descendents throughout the twenty-first century. There is presently a strong consensus among the professional scientists involved in this field that global warming is occurring and that much of it is being caused by the activities of man. These scientists have also provided some guidelines as to what should be done in order to minimize the detrimental effects of these changes in our climate. According to most polls taken in the last couple of years, a slim majority of the general public of our country also seems to believe that man-caused global warming is occurring.

At the same time, however, the public's perceived importance of the issue of global warming has not yet prompted them and their elected representatives to take an appropriate level of action. Of the large number of problems that face us today, only a small minority of the general public considers global warming to be among our most pressing. According to a poll taken in November 2008, by Canwest News Service of Canada, only a minority of the citizens of eleven different countries, including the U.S., indicated that they would be willing to either spend extra money or lower their standard of living if that was required in order to address global warming. In addition, Gallup's annual update on Americans' attitudes toward the environment in 2010 showed a public that was becoming even less worried about the threat of global warming. This stance seems distinctly unwise in view of the real possibility that continued warming could render all of the other endeavors of man pointless in the longer run. Therefore, it is clear that if appropriate levels of corrective actions are to occur in a timely manner, a much greater level of understanding, conviction, and determination will be required of the general public and its elected officials.

The considerable gap that still exists between scientific and public opinions on the subject of global warming constitutes, at the very least, an enormous waste of our nation's extraordinary talent and investment in science. Nevertheless, we will examine in this book whether or not there are valid reasons why this gap exists. Is it possible, for example, that a very visible and skeptical portion of the public knows more and has better judgment than the recognized scientific communities on this issue? Or could it be true, as some have suggested, that the scientific communities are overstating the urgency of the global warming problem for their own ulterior motives? Or is it true, as many Americans seem to think, that the scientific communities have not yet actually made up their minds on whether or not man is contributing significantly to global warming?

There have been many books written on the subject of global warming to date, and therefore, one might question why the world needs another. Given the paramount importance of this topic, however, it seems that there is still a need for additional books that examine and discuss this issue from a variety of perspectives using language that suits different levels of experience and knowledge. From my own vantage point as a scientist, citizen, and grandparent, I will use language and arguments in this book that I believe are well-suited to a wide range of individuals within the general public. I believe that this book will be particularly helpful to those who already know quite a bit about the subject of global warming but are still confused as to how all the pieces fit together and are undecided with respect to their own "bottom-line" conclusions. While trying to be scientifically correct, my style will be conversational without use of equations and graphs. I will endeavor to provide a background and overview of this topic that will be immediately useful for assessing and assimilating the new information and arguments we are daily exposed to by our public media and in communications with other citizens and our elected representatives. In short, I have attempted here to provide an easy read and, hopefully, a page turner on a subject that is relatively complex but of the greatest importance. Each chapter can be read in any order preferred.

While many different aspects of the science that underlies global warming will be mentioned here, it is beyond the scope of this book to explain all of these in detail. Fortunately, there are several other books that do a much better job of this and can be referred to as needed. For this purpose, I particularly recommend *A Rough Guide to Climate Change* by Robert Henson. For those with scientific backgrounds, I strongly recommend the comprehensive research article entitled "Target Atmospheric Carbon Dioxide: Where Should Humanity Aim" by Hansen *et al.* which is available to the public at *The Open Atmospheric Science Journal*, Volume 2 (2008), pages 217-231. Full references to these and other sources of information associated with each chapter are provided in a bibliography of suggested additional reading at the end of this book. For additional information concerning the science of climate change that is particularly well-matched to the contents of this book, visit the author's website at www.ericgrimsrud.com and select the menu tab, "Short Course on Climate Change." Finally, in order to help the reader keep track of the meaning of various terms and acronyms to be frequently used here, a glossary — including some extended definitions — has been provided after the final chapter.

I will finish this introduction with a few personal comments. When contemplating the serious subject of this book, some might feel guilty about the distinct possibility that they have not done enough during their lifetime to address the problem of global warming, and in all likelihood, have contributed to it. It might be helpful, therefore, for the author to confess that he is as guilty as most of that offense. While I was directly exposed to this environmental problem during my professional career, I did not do nearly enough to address it in my own personal life until very recently. Only after my retirement from full-time employment was I able to sit back, study some more, and then clearly see the great need for far more immediate personal and public action on the specific issue of global warming. Therefore, I am in no position to preach to others about their past deficiencies in this area and will definitely not do that here. My object is simply to share my current thoughts with whomever is willing to listen, in the hopes that we can now begin to get our country to do whatever needs to be done in order to forcefully address the problem of global warming before it's too late.

Another personal comment: a question invariably asked of anyone who participates in the climate change debate is "who is funding your efforts?" My own answer to that question is no one other than myself and my wife, to date. I will admit, however, that I have found the Teacher's Retirement Service of Montana to be an ideal means of making ends meet while writing this book. The TRS of Montana provides me with a check every month, I am free to do whatever I want with it, and I do not even have to file a report in order to get the next one! Concerning any proceeds from sales of my book, over the three years since the production of its first edition those have amounted to less than 10% of the substantial (for us, anyway) funds my wife and I have borne for its production, promotion and distribution. One does not generally do very well in the "self-publishing" business.

I was very pleased, therefore, to recently learn from my publisher, I-Universe, Inc, of Bloomington Indiana, that they now wished to assume the role of a traditional publisher of my book. Thus, for this second edition, they will be bearing all costs associated with its revision, promotion and distribution. For this additional assistance in increasing public exposure to my book, I am most grateful. And if I do better financially under this mode of operation, my wife and I intend to use much of those additional resources for doing more of the same during our retirement years — that is, bridging the gap of knowledge concerning climate change that exists between the American public and their scientists — for the sake of our grandchildren.

CHAPTER 1

Man's Effects on the Earth's Atmosphere

In the newspapers, talk-radio shows, and Web site blogs of America, a large portion of the editorials, opinions, and comments dealing with the subject of global warming continue to argue against the notion that man's activities are having a significant effect on global temperatures. These opinions are often then followed by expressions of gratitude and relief by other readers and listeners. Such sentiments are understandable. We all hope that some of the more dire predictions of global warming do not turn out to be true. Also, we recognize that for many of us it might be exceedingly difficult, extraordinarily costly, and definitely inconvenient to squarely address man's effects on our climate. To accomplish this task on a worldwide basis will require an unprecedented level of international cooperation. Nevertheless, after studying the Earth's atmosphere during my career and making numerous measurements of its changing contents, I am not comforted by the arguments against anthropogenic (man-caused) global warming (AGW). At the very beginning of this book, I would like to relate four reasons — based largely on well-known facts and measurements rather than theory — why I am concerned about several of man's effects on the Earth's atmosphere — including those that are likely to lead to AGW.

Our atmospheric blanket is thin: First, it is important to realize that the atmosphere of the Earth is not nearly as large and robust as it might appear to be when viewed from ground level. The atmosphere is, in fact, a relatively thin layer of gas that clings quite tightly to the surface of the Earth. Its pressure approximately halves with every 3.4 miles of increased altitude, and therefore, at a distance of only 20 miles upward, atmospheric pressure is reduced to about 1 percent of that at sea level. If our entire gaseous atmosphere could somehow be condensed into a liquid or solid form of density equal to that of water, its depth would be only about ten yards. About 80 percent of that mass would be nitrogen, a relatively inert substance. The remaining active ingredients would consist almost entirely of oxygen, along with a host of minor components. While the concentrations of these minor components are very low, several of them play essential roles in ensuring that conditions at the surface of the Earth remain favorable to existing forms of plant and animal life. From this initial discussion of how thin our atmosphere actually is, one can already see that it might not be a good place for dumping some of our garbage.

Man can easily affect the contents of the atmosphere: Because the total mass of the atmosphere is so small and the population of the world has grown so large, the activities of man have definitely affected the contents of our atmosphere, especially during the last century, and there are many convincing examples of this. Among these are various classes of halogenated organic compounds (typically these are hydrocarbon molecules on which the hydrogen atoms have been either partially or completely replaced by fluorine, chlorine, or bromine atoms) that have been produced by man for a variety of commercial purposes.

In the first few decades of the twentieth century, there were no chlorofluorocarbons (CFCs) present in our atmosphere because these compounds are not formed naturally. The large-scale production of CFCs was begun by the DuPont Chemical Company in the 1940s; this was then stopped near the end of the twentieth century after it became clear that these compounds were accumulating in the atmosphere and contributing to the destruction of our protective layer of ozone in the Earth's stratosphere. In the refrigerators and air conditioners manufactured today, CFCs have been replaced primarily by another class of halogenated compounds called hydrofluorocarbons (HFCs). As a result, the concentrations of HFCs have increased rapidly during the last two decades from non-detectable levels in the 1980s. While the HFCs are "ozone-friendly," they are greenhouse gases, and as we will see, are increasingly interfering with another basic function of our natural atmosphere. As will be related in chapter 5, I have been personally involved in the measurements of CFCs, HFCs, and similar classes of compounds in the Earth's atmosphere since they were first recognized to be of environmental importance in 1974.

The concentrations of two of the major greenhouse gases — carbon dioxide and methane — are known to have also increased significantly over the last 160 years, by about 40 percent and 150 percent, respectively. While the present elevated levels of carbon dioxide have been attributed primarily to the combustion of fossil fuels during the Industrial Age of man, the reason for the much larger percentage increase in methane observed over this same time period is not clear. Certain agricultural practices, along with fossil fuel use, have been implicated. Another possibility involves an amplifying feedback mechanism by which global warming causes increased emissions of methane from various natural sources, including wetlands, permafrost, and seabed deposits under the Arctic Ocean. This latter possibility is a particularly grave one in that, if large enough, it could cause what is known as a "runaway greenhouse effect" (elevated temperatures cause the emission of more methane, leading to still higher temperatures, causing more methane emission, leading to ... and so on). When I began working in the field of atmospheric chemistry in 1973 with the Air Pollution Research Group of Washington State University, they had begun monitoring the concentration of methane in the background atmosphere and had found it to be 1.5 parts per million (ppm). Today, methane is approaching a level of 2.0 ppm, an increase of about 30 percent over a period of only 40 years.

Some compounds have very long atmospheric lifetimes: We must be particularly concerned about the effects of man-produced compounds that stay in the atmosphere for a very long time after they are released. Unfortunately, all of the compounds mentioned in the examples above have either *very* long (tens of years) or *extremely* long (hundreds of years) lifetimes once they are released into the atmosphere. This means that any environmental effects of pollutants having *very* long lifetimes (such as the HFCs and methane) will persist for several decades, and any detrimental effects of compounds which have *extremely* long lifetimes (such as the CFCs and carbon dioxide) will persist for several centuries even after their man-caused emissions are entirely stopped. An additional problem associated with the continuous emission of long-lived compounds into the atmosphere is that they will tend to accumulate as they are emitted, and their atmospheric concentrations will increase to much higher levels than will those of most compounds that are removed much more quickly.

When long-lived pollutants interfere with natural atmospheric processes: All of the compounds mentioned in the examples above are known to affect one or more basic functions of the Earth's atmosphere that create life-sustaining conditions on the Earth's surfaces. These essential functions include

    (a) the removal of ultraviolet radiation coming from the sun,

    (b) the continuous cleansing of the atmosphere, and

    (c) the maintenance of surface temperatures warm enough for the support of existing forms of life.

The CFCs are known to interfere with function (a) above by causing the catalytic destruction of ozone in the stratosphere. Methane is known to interfere with function (b) by its reaction with the main oxidizing agent of the troposphere, a trace-level substance called hydroxyl radical. The compounds that can interfere with function (c) and thereby lead to increased global warming include all of the compounds discussed above and a multitude of others. Any molecule having three or more atoms will act as a greenhouse gas by absorbing some of the infrared radiation emitted by the Earth, thereby inhibiting the Earth's attempt to cool itself. Therefore, the temperature of the Earth can be expected to increase with enhanced levels of greenhouse gases (not unlike the temperature increase we experience when we put on a heavier coat).

I have made no attempt here to include all of the factors that might contribute to global warming. For that more complex assessment, many other factors have to be included, along with complex computer models of the total atmosphere. The validity of these models must be checked by their application to known climate changes in both our recent and geologic pasts. Only then can they be reliably applied to the future. Because of the complexity of this task, considerable expertise in this particular field is required in order to produce quantitatively reliable predictions. Nevertheless, there are now numerous laboratories and even individuals with these capabilities, and they are turning out models that consider the effects of all imaginable variables. Simply by consideration of the four basic thoughts I have related here, however, it is not at all surprising to me that these simulations of climate have shown that the warming that has been observed over the last century is due largely to human-caused increases in the greenhouse gases of our atmosphere and that the effects of these changes will become increasingly apparent throughout the twenty-first century.

CHAPTER 2

The Ice Ages and Global Warming

The natural occurrence of previous climate changes and ice ages is often used in arguments against the notion of anthropogenic global warming (AGW). After all, humans were clearly not involved in the warming periods that ended each of the previous ice ages. Therefore, why should they be implicated in the present warming trend? This question is raised so frequently within the public discussions of global warming that it is useful to remind ourselves of the scientific response.

By my use of the term, "ice ages" here I am referring to the oscillations between the glacial and interglacial periods that occurred during the last three million years of the last Major Ice Age which began about 35 million years ago. These oscillations are thought to have been initiated by well-known variations in the Earth's orbit and tilt (called the Milankovitch cycles) as the Earth revolves around the sun. These orbital variations cause small changes in where the solar radiation strikes the Earth. Any temperature changes thereby caused are then additionally amplified by resulting changes in the reflection (called the albedo) of that incoming sunlight and by changes in concentrations of the greenhouse gases.

We also have what is thought to be an accurate record of the temperature changes that have occurred and the atmospheric concentration of carbon dioxide (CO_2) existing in the atmosphere over the last 800,000 years. Both of these determinations have been provided by the analysis of ice core samples collected in Antarctica and Greenland. This record shows that CO_2 levels rose and fell between the limits of 180 to 300 ppm during cool and warm periods, respectively, throughout that entire time period.

Over the last 160 years, however, we also know that CO_2 levels have risen from 280 to 393 ppm. By the end of the twenty-first century, CO_2 levels are sure to exceed 600 ppm at the present rate of fossil fuel consumption. If allowed to occur, that would be twice the highest levels ever reached during the previous 800,000 years. The current rate of CO_2 increase is also unprecedented. During the previous ice ages, it took at least a thousand years for a thirty-ppm change in CO_2 concentration to occur, while CO_2 rose by thirty ppm to the present level in just the last seventeen years!

Therefore, what is happening today appears to be driven by a new and different mechanism not previously observed for at least 800,000 years. Since the current rise in CO_2 levels started with the onset of the Industrial Age, the activities of man are clearly implicated. This conclusion has been additionally reinforced by measurements of the relative abundance of carbon's heavier isotopes — carbon-14 and carbon-13 — in atmospheric CO_2. These isotopic "fingerprints" show that the significant amount of fossil-fuel-derived CO_2 present in today's atmosphere is unprecedented. Furthermore, the carbon-13 measurements show that our present CO_2 overload did not come from the natural emissions of volcanoes and the oceans.

It should be noted that most of the information provided in this and the first chapter has been based on direct physical measurements rather than on theory or computer-generated models. These measurements have resulted from the detailed analysis of the ice core samples collected in Greenland and Antarctica during the last two decades and the direct analysis of Earth's atmosphere during the twentieth and twenty-first centuries. The validity of all of these methods is now very well established. Thus, direct measurements alone clearly indicate that the Earth's atmosphere is presently changing in ways not previously observed for at least 800,000 years — and that those changes are occurring rapidly.

CHAPTER 3

Searching for a Flaw in the Theory of AGW

In observing the public debate concerning the degree to which the activities of man have contributed to the warming of our planet, it became evident to me that these discussions often lack a basic understanding of the scientific principles on which the notion of anthropogenic global warming (AGW) is based. Whether one is arguing for or against AGW, one should at least be aware of the basic concepts as well as the measurements that have given rise to this possibility. In this chapter, I would like to focus on some of the key principles of the theory that underlies the concept of AGW. While measurements can tell us something about the past and present, theory is required, of course, in order to predict what will happen in the future. Theory is also the most controversial aspect of the global warming debate. One sometimes even hears the opinion that the notion of AGW has no scientific basis and is merely a hoax driven by political factors. Therefore, in describing the central aspects of the theory behind AGW here, I will do my best to include and assess all of the potential flaws that have been suggested by the deniers as well as the believers of AGW.

A Simple Qualitative View: To show at the very onset of this chapter that the notion of AGW is not an off-the-wall, nonsensical idea, it is useful to first consider the following common-sense, qualitative view of the issue.

The Earth can be envisioned to contain two very different forms of carbon. We can call one of these "geological carbon" (GC). GC includes relatively "inert" substances such as the fossil fuels (including coal, oil, and natural gas) and various inorganic substances such a calcium carbonate (limestone). These forms of carbon tend to stay put in and on the Earth essentially forever if they are left undisturbed.

We can call the other form of carbon "biological carbon" (BC). BC is in all living plant and animal materials and also includes the CO_2 in our atmosphere and the CO_2 that dissolves in our oceans, lakes and streams. The BC forms of carbon are relatively "active" in that they cycle through the atmosphere, oceans, plants and animals very rapidly on a geological time scale.

Now consider the extent to which man has been changing GC carbon to its BC forms. We began doing this on a significantly large scale in about 1850. At that time, our atmosphere contained a natural level of carbon dioxide of 280 ppm or about 550 gigatons of elemental carbon. Today, our atmosphere contains 393 ppm carbon dioxide or about 800 gigatons of carbon. Since we also know that about 500 gigatons of carbon has been burned since 1850, it is readily seen that about 250 gigatons of carbon or about one-half of the total amount of CO_2 emitted by fossil fuel combustion since 1850 is still in our atmosphere! About one-quarter of that excess man-produced carbon dioxide has been found cycling through the surface layers of the oceans of the world and the remaining quarter might be assumed to be cycling though our plants and soils.

The above observed facts clearly demonstrate man's rapid rate of GC to BC conversion and the tremendous effect that has had on our atmosphere. Note that these facts also demonstrate the very slow rate at which the excess BC carbon is removed from the BC cycle. We have seen that about half of the carbon dioxide added by man simply "piles up" or accumulates within our atmosphere. This is expected. It takes many millions of years to naturally convert plant material to the fossil fuels. Another means of BC to GC conversion is called the "weathering" of CO_2 by which the CO_2 dissolved in rain drops or in the oceans comes in contact with rocks that contain inorganic oxides, such as calcium oxide (CaO). A small portion of that dissolved CO_2 will then be converted to limestone ($CaCO_3$). This process is also very slow, however, and has a significant effect only over a time scale of several millennia. We also know that a small fraction of BC descends to the bottom of the oceans each year but, unfortunately, that process is also very slow on a time scale of human relevance. A more rapid process by which some carbon dioxide could be removed from the atmosphere is by its sequestration into the plants, trees, and soils of the Earth. Unfortunately, our planet has not moved in that direction and has undergone much more deforestation than forestation in recent centuries.

OK, but why does it matter if we are adding more carbon dioxide to our atmosphere? The answer to this question is related to how the Earth cools itself. It does this by the emission of infrared radiation (heat) from its surface and from its atmosphere into outer space. Without this cooling mechanism, the temperature of the Earth would continuously rise to extremely high levels. We also know that carbon dioxide is a major greenhouse gas that will absorb a portion of that infrared radiation and thereby work against the Earth's cooling mechanism. Thus, just as you get warmer when you put on a heavier coat, the Earth is expected to get warmer as more carbon dioxide and other greenhouse gasses are added to its atmosphere.

The above common-sense, qualitative ideas give us ample reason to look carefully into the possibility of AGW. To dismiss this possibility out of hand would be foolish.

Arrhenius's Theory: The central quantitative aspects of the theory behind AGW have been known for more than 100 years. Starting in 1896, the preeminent Swedish scientist Svante Arrhenius began to suspect that the extra carbon dioxide resulting from the combustion of fossil fuels would lead to a steady increase in the atmospheric concentration of CO_2, and due to the greenhouse effect of this molecule, this would cause a continuous increase in the temperature of the Earth.

Using the primitive calculation tools of his time, Arrhenius also estimated that a doubling of the atmospheric level of CO_2 over the Industrial Age would cause an increase in global average temperature of about nine to eleven degrees Fahrenheit. This estimate is not so far from the recent prediction of an increase of 3.6 to 8.1 °F for CO_2 doubling reported in 2007 by the United Nations Intergovernmental Panel on Climate Control (IPCC) based on the latest computer models of the Earth's climate.

It is also interesting to note that Arrhenius thought it would take about 3,000 years to reach this point of CO_2 doubling. Living in Sweden as he did, this prediction was a bit of a disappointment to him, in that he concluded he would not live nearly long enough to enjoy a more temperate climate. Arrhenius's error in predicting the CO_2 doubling time resulted by underestimating the amount of fossil fuel that would be burned during the twentieth century and by greatly overestimating the extent to which CO_2 would be absorbed by the oceans. We now know that this CO_2 doubling point could easily be reached within a few decades.

Now let's systematically explore potential reasons why the basic theory of Arrhenius might be wrong. First, we should ask, **Have CO₂ levels in the atmosphere increased as a result of man's activities during the Industrial Age** (as Arrhenius predicted)? This question has already been clearly answered in the affirmative by the direct physical measurements that were described in the previous chapter.

Again, from the analysis of Antarctic ice core samples, we now know that for 2,000 years prior to the beginning of the Industrial Age in about 1850, the concentration of CO_2 in the atmosphere was constant at about 280 ppm. Then, in about 1850, a steady increase in the level of CO_2 began to occur. The present concentration of CO_2 in our atmosphere is 393 ppm, about 40 percent greater than it was at the beginning of the Industrial Age. In addition, CO_2 is currently rising at the unprecedented rate of about two ppm per year. This is about 100 times faster than ever observed during the 800,000-year period of the ice core record!

Okay, so Arrhenius's theory is holding up so far. But now the second part of his prediction needs to be addressed: **Will this steady increase in the level of atmospheric CO₂ cause a continuous warming of the atmosphere?** Alternatively, could there not be some additional phenomena that Arrhenius might not have been aware of in 1896 that might tend to negate a prediction of warming? From my search of both the peer-reviewed literature and the abundance of publications and Web sites produced by the skeptics of AGW, I have come up with a long list of twenty potential reasons why predictions of future warming by CO_2 could possibly be either wrong or of little importance. Each of these is individually considered and discussed below.

- **"But our weather patterns are so variable that we cannot even predict next month's weather much less that in future decades."** It is true that we cannot predict the weather that we will have in any given place next month or next year. But our goal here is to predict average changes in climate, not details of our weather. Climate is the average of weather observed over the entire planet over a given period of time. The measurement and prediction of an average annual temperature is a valid and meaningful indication of climate change.

- **"But even our climate is determined by so many different and unknown factors, that it is hopeless to try to understand and predict it."** This statement is not correct. There are just three basic factors that determine the average temperature of the Earth. One is the intensity of the sun's radiation at our position in the solar system. Another is the fraction of that incoming sunlight that is reflected back into outer space (the scientific term for this is the "albedo" of the Earth). The third basic factor is what's called the "greenhouse effect" by which the Earth's attempt to cool itself via its emission of infrared radiation (commonly referred to as "heat") into outer space is interfered with by both clouds and an assortment of "greenhouse gases". That's it. Those three are the only factors that have a significant effect on the Earth's temperature. Therefore, the task of developing an understanding of Earth's climate is reduced to that of determining what effects, if any, various events occurring on Earth might have on each of these three basic factors. That task is manageable and has been continuously addressed by scientists for more than a century.

- **"But CO_2 is not really a powerful greenhouse gas (GHG)."** This statement is clearly incorrect. Carbon dioxide definitely absorbs infrared radiation and is currently the second-strongest absorber of that radiation emitted by the Earth.

- **"But water is the most potent GHG, and therefore, the increasing levels of CO_2 will only be of secondary importance."** The latter half of this statement is not only also incorrect, but the opposite is actually closer to the truth. This is because water vapor is not a "forcing agent" for warming like CO_2 and the other long-lived GHGs. Instead, water serves primarily as a "feedback gas," in that it amplifies any initial warming caused by the long-lived GHGs. The amount of water present in the atmosphere quickly increases or decreases with changes in temperature. Therefore, any small increase in temperature caused by CO_2 and the other GHGs causes more water to be evaporated and adds to the original warming effect of CO_2. Conversely, as the temperature is decreased for any reason, water vapor rapidly condenses out of the atmosphere as rain or snow. Most of the other important GHGs, including excess CO_2, stay in and are well-mixed throughout the entire atmosphere for many years after they are emitted. It is interesting to note that Svante Arrhenius was aware of this important factor back in 1896 and included it in his early models.

- **"But in the previous warming periods that followed each ice age, an increase in the CO_2 levels always lagged behind, rather than preceded, an increase in temperature."** This statement is true, and in addition to those discussed in chapter 2, provides another reason for why we believe that the primary cause of the warming we are experiencing today is fundamentally different from that which occurred after each ice age. The previous ice ages are thought to have been initiated and ended by subtle changes in the Earth's position relative to the sun (Milankovitch cycles). As the Earth was thereby warmed after an ice age, progressively more CO_2 was driven by that heat from the oceans into the atmosphere. This extra CO_2 then served to amplify the warming trend because of its greenhouse properties. Later, when further changes in the Milankovich cycles began to cool the Earth and drive it toward another ice age, a continuous decrease in CO_2 would have occurred due to its increased absorption by the cooler oceans. Therefore, the changes in the CO_2 level that occurred during these warming and cooling periods lagged behind the changes in temperature.

At the present time, however, about 7 billion people live on our planet and a great deal of extra CO_2 is being emitted into the atmosphere via the combustion of fossil fuels. Unlike all of the previous warming periods discussed above, we are now observing a steady and rapid increase in the level of CO_2 that has been preceding — rather than following — a change in temperature. The fact that we have never observed this before during the 800,000-year record of the Antarctic ice cores provides additional strong evidence that what is happening today is due to the activities of man and is not of natural origin.

- **"But with time, other GHGs might become more important than CO_2."** Yes, this could possibly happen. The percentage increases in methane and the man-made halogenated compounds, for example, have been much greater in recent decades than that of CO_2. However, the warming effects of these gases will only add to the warming caused by CO_2, thus increasing the magnitude of the problem. In addition, the excess carbon dioxide merits greater concern than the other GHG's because of its extraordinarily long life time.

- **"But the cooling effect of increased cloud formation will neutralize the warming effect of the CO_2."** It is true that enhanced levels of water vapor (due to increased temperature) and an associated increase in cloud formation will have a cooling effect by increasing the reflection of incoming sunlight (the albedo effect).

However, these clouds will also have a warming effect by absorbing the infrared radiation (heat) emitted from the Earth. In addition, during the evening, only the warming effect will be operative. Both of these feedback effects will also depend on the type and elevation of each cloud. Low clouds tend to reflect sunlight more effectively than high clouds. The net effect of clouds alone on warming is therefore thought to be somewhere near neutral (i.e., no or little effect). While this issue is presently one of the least-well-understood aspects of climate change, there is no reason to expect that it will fortuitously offset the warming effect of the GHGs.

- **"But 'global dimming' by particulate matter will offset warming by the GHGs."** It is true that small particles (also called aerosols) injected into the atmosphere can lead to a significant cooling of the Earth due to the increased reflection of incoming solar radiation they cause. Evidence for this is provided whenever a large volcano punches particulate matter high into the Earth's stratosphere, where it can take a few years to dissipate. The eruption of Mount Pinatubo in the Philippines in 1991 provided a clear example of this.

  This dimming effect is also caused by the ground-level emission of particulate matter from a wide range of man-controlled sources, including coal-fired power plants (and especially those plants that do not have state-of-the-art scrubbing systems on their smokestacks). This effect is coupled to that of clouds discussed above because particulate matter is known to facilitate the formation of clouds and prolong their existence. Since particulates from these sources are injected into the lower region of the atmosphere (called the troposphere), they are removed by natural processes within a few days.

It is believed that aerosols have, in fact, offset some of the warming caused by the GHGs. Climate models over-predict the magnitude of warming observed over the last century when the effects of aerosols are not included. The problem with the argument that aerosols will completely offset GHG warming is that aerosol emissions from human activities are more likely to decrease than increase in the future, in order to avoid other environmental and human health issues associated with particulate matter. Since current emissions of aerosols have only partially offset GHG warming, we can't rely on such emissions to mitigate climate change if GHG levels aren't stabilized or reduced.

- **"But changes in the Earth's climate are affected by other and potentially more dominant natural factors."** Of course, this is true. The greenhouse effect is only one of several natural factors that have affected our climate. Other factors include variations in the intensity of the sun, continental drift, variations in the Earth's orbit around the sun (the Milankovitch cycles), asteroid impacts, and periodic eruptions of volcanoes. We know a great deal about all of these natural factors, however, and none of them can explain the rapid warming that has occurred in recent decades. (Note:

Variations in the intensity of the sun will be further discussed below.) In addition, we also know that the strong warming effect of increased CO_2 and the other GHGs will be superimposed on any effects that might possibly be caused in the future by these other natural factors. Since the global trend has not been toward lower temperatures during the Industrial Age, it is clear that a dominant natural factor causing cooling has not been operative over this period. And again, it should be recalled that the warming effect of excess carbon dioxide will last essentially "forever" on a time scale of relevance to advanced human civilizations.

• **"But the Earth is so massive; some extra CO_2 molecules produced by man couldn't possibly affect our planet."** This statement is clearly incorrect. While the Earth is massive, its atmosphere is not. Within this relatively thin film of gas surrounding the Earth, some of its minor components (including CO_2) play central roles in creating conditions that are favorable to existing forms of life.

• **"But the oceans of the world will absorb most of the extra CO_2."** This is what Svante Arrhenius thought and is one of the reasons why he mistakenly predicted that it would take about 3,000 years for the level of atmospheric CO_2 to double. It turns out, however, that the exchange of CO_2 between the atmosphere and oceans occurs much more slowly and is more complicated than he initially thought. First, the relatively thin surface layers of the oceans mix exceedingly slowly with the colder depths beneath them, and many hundreds of years are required for complete mixing to occur throughout the oceans. Therefore, the surface layers become saturated relatively quickly and absorb much less CO_2 than they would if vertical mixing of the oceans was fast.

In addition, as the temperature of the oceans slowly increases due to global warming, we know that the ability of the oceans to absorb atmospheric CO_2 will be continuously diminished. The absorption of CO_2 by the oceans also causes the acidity of the surface layers to increase. This also diminishes the amount of CO_2 that the saturated surface layers can absorb (due to a shift in the acid-base equilibria of the oceans). Therefore, these two factors mean that our oceans will progressively become less able to absorb the CO_2 emitted into the atmosphere by both natural and anthropogenic processes as the temperature and acidity of the oceans increase.

It is also now clear that the increased acidity of our oceans caused by the enhanced levels of CO_2 in our atmosphere is having detrimental effects on the ecosystems of our oceans. For example, we have already noted a great deal of damage suffered by the coral reefs of our planet due to the increased acidity of the oceans' surface layers. Clearly, organisms living in the oceans do not want the CO_2 overload any more than those living on land.

• **"But doubling or even tripling the level of CO_2 won't** matter because the infrared radiation it absorbs is already completely absorbed by the present level of CO_2." A common analogy of this argument is still in wide circulation today, that of a window painted black. Would a second coat of paint applied to that window make a substantial difference in the amount of light that passes through it? The answer to this question, of course, is no, thereby supporting the claim made above. The passage of infrared radiation through our atmosphere differs substantially, however, from that of the window painted black. This is primarily because the greenhouse gases in the atmosphere emit infrared radiation as well as absorb it. Thus, the atmosphere must be viewed as consisting of many individual layers, from the Earth's surface to its outer edges, with each of these layers absorbing and emitting infrared radiation in all directions including downward back towards the Earth's surface. In addition, each of these layers differs greatly with respect to its pressure, temperature, and water vapor content. Since the concentration of water vapor decreases sharply with increased altitude, carbon dioxide becomes the major greenhouse gas at high altitudes where there is almost no water. For this system, the heat retention provided by our current atmospheric CO_2 level is expected to be continuously increased by the addition of more CO_2.

- **"But we shouldn't rely only on theoretic models of climate. We also need physical, observable evidence of the likely effects of increased CO_2 levels on climate. I** agree entirely with this statement and, fortunately, we have an abundance of such evidence. For example, from the Antarctic and Greenland ice core records of the last 800,000 years, we know what the temperatures of the Earth and the atmospheric CO_2 levels were over that entire period and therefore know what past relationships between CO_2 changes and temperature have been. This relationship consistently turns out to be a temperature increase of about 11 degrees Fahrenheit for a doubling of CO_2 concentration under the conditions of the glacial and interglacial periods (see Hansen *et al*, 2008). This level of temperature sensitivity is expected to apply to our present condition in which the Earth also has a great deal of ice on it. Other measurements obtained from ocean bottom core samples have shown what the corresponding temperature sensitivity to CO_2 changes was when the Earth had no ice on it — as was the case prior to about 35 million years ago. Along with other geological information concerning that "water-world" era, we know that the Earth's temperature sensitivity to CO_2 changes was then about 5 degrees Fahrenheit or approximately half that observed later when the Earth had lots of land ice. From these measurements we have learned that the Earth's sensitivity to CO_2 changes consists of two parts, one of these is equal to about 5 degrees and includes what we call the "fast feedbacks" that are associated with relatively rapid changes in water vapor, clouds and sea ice. The other component is equal to about 6 degrees and includes the "slow feedback" effects of the slower changes in land ice and the land itself. Both of these effects are of importance when the Earth has a lot of ice on it, as it does today, leading to a total sensitivity of about 11 degrees Fahrenheit. Thus, the record that Mother Nature has left in the ice and ocean bottom core samples indicates that a doubling of the CO_2 level will cause a temperature increase of about 5 degrees Fahrenheit in the shorter term (over the next several decades) and about 11 degrees over the next couple of centuries. This is, indeed, an alarming amount of temperature change and the important point I am making here is that we already know this from physical measurements of the past rather than theory. The role of theoretical models is to add additional detailed understanding to those measurements, not to override them. There is a golden rule in science that goes like this: "theory guides, but experiment decides".

• **"But the current climate models are flawed and their predictions of AGW are probably not correct."** It is true that the models are not yet good enough to reliably predict all aspects of climate (such as "the Midwest is having an unusually mild winter this year"). Nevertheless, they are entirely adequate for predicting basic aspects of long-term climate changes, such as the effects of the GHGs on average global temperatures over a period of several decades.

• **"But temperatures have not been consistently rising since the beginning of the Industrial Age."** Since the beginning of the Industrial Age in about

1850, the average temperature of the Earth has risen about 1.6 °F. This temperature rise has not occurred continuously and smoothly, however. From one year to the next, other factors can also affect temperature. The most important of these is thought to be the "global dimming" effect of particulate matter discussed above. If caused by an unusually large volcano, this emission into the stratosphere can cause a cooling effect that lasts for a few years after the event, as did occur following the eruption of Mount Pinatubo in 1991. In addition, significant global dimming events have been attributed to prolonged activities of man, such as World War II and the continuous emissions of power plants that did not have state-of-the-art scrubber systems on their smokestacks. Without these forces for cooling during the Industrial Age, the Earth would have warmed even more than it did.

- **"But a temperature change of only +1.6 °F shouldn't do any harm."** While this amount of temperature increase (observed over the Industrial Age) is indeed very small relative to the temperature fluctuations we experience every day, changes of this magnitude in the average temperature of the Earth's near-surface air and oceans are in fact very significant and are expected to affect our climate.

  For some perspective on this point, consider the fact that the changes in average temperature that occurred between the coldest periods of the previous ice ages and the warm periods in between was only about +10 °F —

and that those changes took about 5,000 years to occur. Today, the temperature of our planet is beginning to increase beyond the comfortable warm period we have enjoyed throughout the 10,000-year period of human civilization and is expected to go well past the 1.6 °F increase we have already observed during the Industrial Age. In its 2007 report, the IPCC estimated that the increase in temperature caused by a doubling of the pre-Industrial Age level of CO_2 from 280 to 560 ppm will likely be between 3.6 and 8.1 °F. In statistical language, use of the word "likely" indicates a 66 percent probability of this outcome (the greatest cause of the uncertainty in this prediction is related to cloud behavior). The majority of models have predicted a most probable temperature increase of about 5 °F (note that 5 °F is also the value for short-term fast-feedback sensitivity deduced from the ice and ocean bottom core records). At the present rate of CO_2 emission, this Industrial Age doubling point will be reached well before the end of the twenty-first century. A 5 degree F increase by the year 2100 would make that year the hottest in more than a million years! Essentially, all models of the Earth's climate suggest that this outcome could have catastrophic consequences.

- **"But global cooling occurred during the last year."** This is a statement that was frequently heard after the year 2008 and is one that we might again hear after 2011. While the statement is correct, it has no long-term significance. A period of one year is much too short to deduce a trend in global warming or cooling. While the year 2008 was, indeed, the coolest year up to that point in the twenty-first century, it was also the tenth hottest on record. And in both of the subsequent years of 2009 and 2010, the global average temperature rose again to the highest levels ever recorded in the 20th and 21st centuries. In 2011, it dropped slightly again as it did in 2008.

Thus, it appears that the slight cooling that occurred in the recent years of 2008 and 2011 represents no more than statistically expected annual variations that are superimposed on an overall long-term warming trend that is apparent only when viewed on a decadal time scale (see the World Metrological Organization's statement concerning the year 2011 on the Web at: www.wmo.int/pages/mediacentre/press_releases/gcs_2011_en.html)

- **"But the intensity of the sun changes with time, and that has possibly been the major cause of recent warming."** It is true that changes in the intensity of the sun do occur. Throughout the 10 billion-year lifetime of its initial bright stage, the sun's intensity continuously increases. The intensity of the sun is now thought to be about 25 percent stronger than it was during its infancy, about 4.5 billion years ago. During the very short period of time from the beginning of the Industrial Age to the present, however, that long-term intensity increase has been so small as to be undetectable. On top of that gradual increase, a repetitive eleven-year cycle of intensity increase and decrease is caused by the varying frequency of sunspots. The magnitude of this variation has been less than 0.1 percent of the total intensity however, and cyclic variations in the Earth's temperature of an eleven-year frequency have not been observed. In addition, during the first half of the twentieth century, a gradual increase in total solar intensity of about 0.1 percent was observed. While this change might have contributed to a small increase in temperature over that fifty-year period, it would have had no effect after 1950. Therefore, the relatively large increases in temperature that have been detected in the last several decades cannot be explained by any of these known natural variations in solar intensity.

- **"But it will take a long time before AGW becomes serious. By then, we will have figured out a way to solve the problem."** While it would indeed be convenient if this statement were true, for several reasons, it does not appear that it will be. First, given the present rate of CO_2 increase in the atmosphere (about 2 ppm per year) and with the prospect of emerging countries rapidly building new coal-fired power plants, we are presently on a course that will increase the level of CO_2 in the atmosphere to about 500 ppm by the middle of this century! Most models of climate change suggest that a CO_2 level of that magnitude might have catastrophic consequences on life as we know it (the last time the atmosphere contained 500 ppm CO_2 was more than 35 million years ago when there was no ice on our planet and sea levels were about 70 meters higher than today).

We stand a much better chance of avoiding some of the direst outcomes of AGW if the CO_2 level is not allowed to exceed 400 ppm and then decreases steadily. Note, however, that the 400 ppm level is almost certain to be reached **in just four years** at present emission rates.

In addition, it has recently been shown (Soloman et al., *Proceedings of the National Academy of Science*, vol. 106, Feb. 10, 2009, which can be seen at the Web site *www.noaanews.noaa.gov/stories2009/20090126_climate. html*) that if and when we do manage to completely stop anthropogenic emissions of CO_2 the cooling due to the subsequent slow decrease in atmospheric CO_2 will be offset by progressively slower heat loss to the oceans. As a result, the warming effect caused by the amount of atmospheric CO_2 we have generated at any future point in time is expected to be with us for several millennia after that point! This prediction has an uncomfortably high probability of occurrence and is based on our rapidly growing knowledge of the processes that control CO_2 and heat exchange between the atmosphere and oceans. In short, this means that the increased heating we cause each day by our continued emissions of CO_2 is irreversible and will be with us forever (that is, if we define the words "irreversible" and "forever" to mean something that lasts for a few millennia).

It is also important to note that the above forecast of our future does not include the less predictable (but even more dangerous) possibilities of reaching various "tipping points" after which "runaway" feedback processes could lead to even greater irreversible increases in temperature. For all of these reasons described in this section, our continued additions of anthropogenic CO_2 to our present atmosphere greatly worsens our future prospects and must be reduced substantially starting now.

- **"But scientists cannot be trusted. They tend to falsify their observations and skew their conclusions in directions that will serve their own selfish interests."** I spent my entire career working with professional scientists and noted that they do indeed possess the same human limitations and foibles that all groups of human beings have. So, of course, they do make mistakes, usually unintentional but sometimes in line with their personal preferences. Nevertheless, I also have the highest regard for our scientific communities and the processes they use to assess the quality of work reported by scientists. Those processes recognize their human limitations and are designed to find and correct errors of both measurement and assessment so that the best work and thoughts can float to the top. This is generally accomplished via an extensive peer-review system in which the work of all professional scientists can be read, studied, and repeated by others. In the ensuing discussions that appear in subsequent peer-reviewed papers and conferences, the best science usually emerges even if it takes a while for it to reach a useful level of maturity. From there, minor additions to that mature view often follow and occasionally landmark new discoveries will overturn either part of or the total applecart. That's how science works and the resulting opinions that then emerge are very likely to be the best available anywhere. For centuries now this has proved to be an excellent system and explains why, for example, we believe today that the Laws of Gravity correctly describe the attraction between two massive objects. We do not believe these things simply because Sir Isaac Newton might have been a nice or credible guy. Similarly, the field of atmospheric science has now existed for more than a century and the basic tenets of Svante Arrhenius's model for carbon dioxide's effect on temperature have held up to continuous scientific investigation ever since he proposed them in 1896.

Looking for additional arguments against the theory: In looking for reasons to question the theory behind AGW, I thought I might find some that were particularly compelling in the numerous anti-AGW articles that have appeared in books, newspapers, Web site blogs, and other publications throughout the last two decades. One would expect, after all, that the very best arguments against the notion of AGW would have risen to the top. Upon inspecting these information sources, however, I found little relevant information in addition to the general types of arguments that I have listed and discussed above. In the anti-AGW articles I have read, a common pattern seems to be to avoid careful consideration of the main reasons for why we think AGW is occurring, and instead, to focus on other points of secondary or even no importance that might be less well understood.

For example, let's consider just one of these side issues often referred to by the deniers of AGW. In 2007, Svensmark and Calder published a book entitled *The Chilling Stars: A New Theory of Climate*

Change in which a previously unrecognized effect on climate was suggested to be more important than that of the greenhouse effect. According to this theory, cloud formation on Earth might be influenced by cosmic radiation (mainly high energy protons coming from outer space) and the amount of this cosmic radiation striking the Earth might be affected by sunspots and solar flares of the sun. Since there have been no supportive correlations noted in recent decades between sun spots and either cloud formation or the Earth's temperature, this theory has not yet received much support. Nevertheless, it is also true that many aspects of sunspots and solar flares are not yet well understood. The same can be said of interactions of solar flares with cosmic radiation and the significance of cosmic radiation on cloud formation. Therefore, some skeptics of AGW have insisted that we delay any corrective action concerning AGW by the GHGs until all of the points associated with this new theory are better understood. This type of argument takes the focus off the central issue of warming by the greenhouse gases. Whether or not any new insights into the origin, nature and effects of sunspots are found in the future, that insight is most unlikely to have any effect on the amount of warming we now know that we will experience from increased levels of greenhouse gases.

So do we know enough? We presently know a great deal about AGW and certainly will be learning more every year. As the world continues to burn increasing amounts of fossil fuels, however, there are some very good reasons why we should decide sooner rather than later how we are going to respond to the CO_2 question. One of these is that the excess CO_2 resulting from the combustion of fossil fuels has an extremely long effective lifetime (several centuries to a few millennia) once it is emitted into the atmosphere. Another is that the full heating effect of the CO_2 that we will have accumulated in our atmosphere at any future point in time will require several centuries to overcome the thermal inertia of the Earth. While this latter effect might be considered good news today in that it delays the rate of warming, it works both ways and will also delay any changes back to previous conditions should we manage to make those changes at some time in the future. Therefore, what we do today and in the next decade will affect the atmosphere and the Earth's climate for decades, centuries, and even millennia to follow. Thus, on the time scale of Western civilization, the additional climatic changes we cause each day will last essentially "forever." Every day we proceed with business as usual, we are making the climate less habitable for essentially all future generations.

In deciding between the various options available to us, it would be most helpful, of course, if we knew for sure whether or not all details within the theory of AGW are absolutely correct beyond all levels of doubt. Unfortunately, no scientific theory can ever be proved to be absolutely correct. Even if the entire American Midwest turns into a dust bowl by the middle of this century, that outcome would still not prove that the theory of AGW is absolutely correct and there still would undoubtedly be individuals who claimed that other factors caused that disastrous extent of warming. Scientific theories can only be proved to be wrong. That's the way science works. All we can do is continuously test and retest theories in order to determine whether they hold up to these tests or not. If they do hold up, the theory is considered to be viable, and if they do not, the theory is considered to be flawed. This is exactly what we have attempted to do here.

In our search for a reason why the basic theory of AGW is flawed, however, no such reason has been found. In fact, the basic theory holds up very well against the potential flaws that I have been able to find in my inspection of all related sources of information. This outcome is not surprising, of course, because the hypothesis proposed by Arrhenius in 1896 — denied by almost every expert during the first half of the twentieth century and steadily advancing during the second half — is now as well accepted as any scientific proposal of its nature can be. Therefore, whether we like it or not, it appears that the only responsible thing to do at this late date is to admit that existing theory behind AGW is extremely likely to be correct and immediately face that prospect as forcefully as we can. The only scientific argument these days against that course of action seems to be "but we don't know everything yet" and since that argument will always be true, it is a very poor one.

CHAPTER 4

Are Scientists Really Divided on the Issue of AGW?

A major portion of the general public, including the majority of those with whom I have personally spoken, have the impression that "even the scientists are divided" on the subject of AGW. Therefore, a large portion of the public quite understandably feels that until the scientists make up their minds on this issue, why should they believe that AGW is occurring? As a result of this, a critically important point in the climate change controversy today is whether or not it is really true that "even the scientists are divided" on the subject of AGW. This simple statement will be carefully examined in this chapter.

For starters, let's first consider the documented fact that the Intergovernmental Panel on Climate Change (IPCC) of the United Nations and almost all of the leading scientific organizations of our country, including the National Science Academies, the American Meteorological Society, the National Research Council, the American Geophysical Union, the Geological Society of America, the American Chemical Society, the American Physical Society, and the American

Association of State Climatologists, have stated unequivocally their view that AGW is rapidly occurring and should be addressed in a timely manner. The views expressed by each of these organizations result from thousands of scientific studies and observations carried out over many years by scientists who have achieved the highest levels of education in their respective fields and have remained active in their research areas up to the present time. In addition, if one reads the peer-reviewed articles that appear in the scientific journals of our country today, or if one attends the numerous scientific conferences that are presently being held throughout the world, one is no longer likely to read or hear serious scientifically based debates as to whether or not AGW is occurring, as one did during the previous several decades. Instead, scientific discussions are now focused almost entirely on the myriad details of the anticipated climate changes. The high degree of scientific consensus on AGW that I just described has been documented by Naomi Oreskes and reported in *Science,* one of the most highly regarded journals in all of science.

Given the information provided above, what does the statement "even the scientists are divided" really mean? Are there actually two entirely different — but equally valid — views concerning AGW?

There is no doubt that there are still a few appropriately qualified scientists who do not yet accept the notion of AGW. This fact is not so much surprising, however, as it is inevitable. Just as in all other areas of science, the consensus view on any topic of some complexity is never supported by all of its qualified members. If we required unanimous agreement on all complex physical systems in science, no issue would ever be defined to be "settled." Therefore, whenever the statement "even the scientists are divided" is being used in this context, it means essentially nothing; it would always be true for all complex issues. The only point of practical relevance is whether or not there is a strong scientific consensus on a given issue, and there clearly is that level of agreement on the topic of AGW. If a minority view of significant size existed within any of the scientific fields involved, a minority report would have also been issued by that scientific organization.

A better understanding of the questions posed above is obtained when we acknowledge the effective activities of some special-interest groups whose financial interests could be damaged by the message being provided by our national scientific organizations. A common strategy of some of these well-organized and exceedingly well-financed groups has been to convince the public that "even the scientists are divided" by creating their own version of a "scientific community" whose opinion is certain to be more in line with their own preferences.

This strategy is not new, of course, and has been used many times before by various other interest groups that have found themselves to be at odds with the best science available on their topic of concern. For example, we have also seen "pseudo-scientific experts" of this ilk attack the legitimate scientific consensus behind the issues of acid rain, stratospheric ozone depletion, and secondhand smoke. This strategy has been quite successful in influencing public opinion, because many within the general public are not well equipped for distinguishing between an actual scientific expert in a given field and a scientist who is simply very good at appearing to be one. In addition, the bottom-line message of these pseudo-experts concerning AGW is invariably more comforting than that of the scientific community, and therefore, is much easier to "sell" to the general public.

In my home state of Montana we have recently seen an excellent example of this. A gentleman named H. Leighton Steward has presented himself to Montanans as a scientific expert on the subject of climate change. Mr. Steward was, in fact, the Director of EOG, a gas and oil company formerly known as Enron. He also has been the spokesperson for a fossil-fuel advocacy group called *Plants need* CO_2, whose advertisements have been shown frequently throughout Montana in recent years. On June 9, 2010, he provided the citizens of Billings and the students of MSU Billings with a presentation sponsored by the Montana Petroleum Association called *Our atmosphere needs more* CO_2. For your inspection, his basic message can be seen in the 2010 issue of the *Montana Treasure State Journal*, the official publication of the Montana Petroleum Association, pages 28-32 (see it on the Web at: montanapetroleum.org/assets/PDF/articlesReports/2010-Treasure-State-Journal.pdf.) In his presentations, Mr. Steward assures us that the Earth's temperature can increase only by a very small and insignificant amount — even if we let carbon dioxide levels increase without constraints during the rest of this century and into the next — allegedly because the present level of carbon dioxide already absorbs essentially all of the infrared radiation that carbon dioxide can absorb. That is, he claims that carbon dioxide's effect on warming is already saturated. You might recognize that this is nothing more than the old "window painted black" analogy that was previously discussed and dismissed in Chapter 3. Mr. Steward's comforting model is of no relevance to the real world because it accounts only for the absorption of infrared radiation by the greenhouse gases and does not include the simultaneous emissions of this radiation by these same molecules throughout the atmosphere. While all professional climate scientists, including Svante Arrhenius way back in 1896, know that the emissions of infrared radiation by the greenhouse gases are as important as their absorptions, apparently Mr. Steward does not. In any case, the Montana Petroleum Association prefers Mr. Steward's version of "happy science" no matter how childish it is and does its best to sell it to the public of Montana. I have personally attempted to help the Montana Petroleum Association with their misconceptions and their misrepresentations of climate science to the people of Montana, but they have shown no interest in updating their knowledge beyond that described above. Neither did Mr. Steward when I personally called this matter to his attention.

A common method of attack by the special interests is to focus on and misrepresent the "uncertainty" of conclusions associated with any environmental issue. Professional scientists know that the conclusions drawn from scientific studies of complex systems must always be accompanied by estimates of the uncertainties of those conclusions. In addition, scientists know that absolute certainty is never possible and that the criteria of knowing everything "for sure" should never be required in order to take corrective action on complex issues. A large portion of the lay public does not understand the mathematical language of statistics, however, and by a clever but incorrect argument, can be led to believe that even statistically solid scientific conclusions are either suspect or tentative.

Another method commonly used to create the impression that "even the scientists are divided" is to provide the public with a long list of supposedly "qualified scientists" whose opinions are the opposite of those expressed by our national scientific organizations. An example of this is the Global Warming Petition Project that is frequently offered as "proof" that the scientific community is divided. The Web site of the Petition Project (www.oism.org/pproject) includes a list of more than 31,000 scientists who allegedly do not believe that man is contributing significantly to global warming.

First, it should be noted that the Petition Project was started about fifteen years ago in response to the Kyoto Treaty of 1997. In view of the fact that the most convincing physical evidence of AGW has been observed and reported during the last decade, it is not at all clear that its signers would choose to sign it today. In addition, the only requirement for signing the petition is an undergraduate degree in any area of science — and even that minimum requirement did not require verification.

Upon inspecting the Petition Project's list, the first thing that struck me was how few of the names I recognized as being participants in the field of atmospheric science. This is a point of some significance to me because I spent a major portion of my own career working in the field of atmospheric chemistry and learned the names of many experts in that field. Another telling point is that almost all of the names I did recognize were individuals whose research area had little or nothing to do with the fields of either atmospheric science or climate change. Why, I asked myself, were these people invited to sign this document concerning a subject to which they had no visible connection? The general field of science is so enormous and includes so many sub-disciplines that scientists in one field usually have only a superficial level of knowledge and understanding of the others.

Another document of this general type should be mentioned here because it appears to have been provided by use of taxpayers' money. I am now referring to a staff report (Inhofe) of the U.S. Senate on Environmental and Public Works Committee issued on December 11, 2008. The title of this report was *U.S. Senate Minority Report: More than 650 International Scientists Dissent Over Man-Made Global Warming Claim*. This document includes statements made by various international scientists whose accomplishments in the area of climate change are not specified, but who are willing to attest to their skepticism concerning AGW. Upon reading the document, a distinctly poor impression is made of its scientific contents. In it, I did not find any substantial arguments that we have not already considered and discounted in chapter 3. Most of the comments offered only the "bottom-line" opinions of these individuals with shallow, if any, justifications. Many comments were absolutely sophomoric, such as the frequent suggestion that CO_2 is not an important greenhouse gas. Many of the comments focused on issues of a political or economic nature, rather than on the science. In short, this report was not based on anything close to the best science available, and certainly does not provide convincing evidence that the real community of scientists who are doing credible research in this area is "divided" on the issue of AGW.

The general public should understand that the scientific community does not — and should not — decide important issues simply by a count of hands among all minimally qualified scientists. After all, is a person who is in need of heart surgery likely to seek the advice of all minimally qualified health care providers or would that person seek the advice of physicians who are both experienced and active in the field of heart surgery?

An additional reason why many members of the general public have the impression that "even the scientists are divided" on the issue of AGW is related to the mass media's inclination to present "both sides" of all issues. Therefore, when the scientific case for AGW is presented, the media often includes or follows up with a counterpoint usually provided by a member of that community of pseudo-experts. If a member of the general public is not well equipped to assess either the scientific credentials of the two presenters or the scientific merit of their arguments, he or she is left only with the impression that the two scientists selected by the media were indeed "undecided" on the issue.

An excellent example of this was recently provided by The *Daily*

Inter Lake, my local newspaper of Kalispell, Montana, in its coverage of Dr. Steven Running's presentation at our local community college in September 2008. Dr. Running, of the University of Montana, is an internationally recognized expert in the field of forest ecology and is one of the co-recipients of the 2006 Nobel Peace Prize. After his thorough presentation of the science behind AGW, numerous questions and comments from the audience were entertained by Dr. Running. One of these comments came from a gentleman, introducing himself as an "atmospheric physicist," who forcefully stated his opinion that AGW is not occurring and that the scientific experts in his field were still very divided on the issue. Dr. Running listened politely and then simply stated his view in no uncertain terms that climate change scientists were not, in fact, significantly divided on the issue of man's involvement. In the ensuing newspaper report covering Dr. Running's presentation, the comments and alleged credentials of this denier of AGW were given considerable coverage along with those of Dr. Running. Thus, by the media's tendency to provide "fair and balanced" news, local readers were once again given the impression that "even the scientists are still divided" on the issue of AGW.

Many members of the general public will undoubtedly continue to deny the notion of AGW, and of course, it is their right to do so. In addition, many might continue to think that they have substantial scientific company in their camp. My advice to these individuals, however, is that they carefully examine the credentials of their chosen experts and also try to determine whose interests they really serve. In our country, it is entirely legal to be an advocate masquerading as a scientific expert, and as far as I know, public admission of one's financial sponsors is not required. As the effects of AGW become increasingly clear to the general public, the pseudo-scientific experts of this type are likely to decrease in number and effectiveness. Hopefully well before then, the American public will come to realize that they are still being subjected to the most dangerous hoax ever perpetrated on the human race and will increasingly rely on their real scientific community for guidance.

CHAPTER 5

*Lessons of the Chlorofluorocarbon/
Stratospheric Ozone Controversy*

While we are presently engaged in an issue of monumental importance related to anthropogenic global warming (AGW), it is instructive to recall events surrounding a previous environmental controversy also concerning the Earth's atmosphere that began in 1974 and continued for about two decades. That controversy concerned the potential effects of a set of man-made compounds called chlorofluorocarbons (CFCs) on the Earth's atmosphere and more specifically on the ozone layer of the Earth's stratosphere. As I was directly involved in the scientific aspects of that issue from its very beginning, I had a ringside seat to ensuing events. I learned a bit about the challenges of getting a serious environmental threat to be recognized and acted upon by the public and the affected industrial interests. In this chapter, I would like to describe some of the basic aspects of that controversy and reflect on the lessons thereby learned.

The chlorofluorocarbons (CFCs): The two most common of these were CFC-11 (CFCl$_3$) and CFC-12 (CF$_2$Cl$_2$). These compounds were first synthesized by the DuPont Chemical Company in 1928 and were being mass produced by the 1940s under the trade name Freon. The CFCs were found to serve exceedingly well in a variety of applications, including propellants in aerosol spray cans, as the recycling fluid in air conditioners and refrigerators, and as the filler gas of polyurethane insulation foams. In addition, they were chemically and biologically inert and completely safe to breathe once released into the air. Because of the great success of CFCs in these applications, their production rate increased exponentially (doubling about every seven years) until 1975. After 1975, the production of CFCs remained roughly constant until 1988 and then gradually decreased until their large-scale production was essentially terminated in 2002.

Theory of stratospheric ozone depletion: The recognition of potential environmental problems often occurs only after the invention of some device that allows the offending chemicals to be "seen" in environmental samples. In the late 1950s, an English scientist named James Lovelock invented such a device, called the Electron Capture Detector (ECD). This device, along with the simultaneously emerging technique of gas chromatography, has revealed the presence of all manners of trace-level substances in environmental samples and is still one of the most sensitive detectors in common use today. It is not a coincidence, therefore, that Rachel Carson's famous book, *Silent Spring*, appeared in 1962, a few years after Dr. Lovelock's invention. The publication of this book brought environmental concerns to an unprecedented level of public attention in the U.S. and led to a nationwide ban of DDT and other pesticides. The environmental movement this book inspired undoubtedly led to the creation of the

Environmental Protection Agency by the Nixon administration in 1970.

In 1971, James Lovelock then applied his ECD to the analysis of the Earth's atmosphere and immediately found CFC-11 and a few other highly chlorinated compounds (including carbon tetrachloride and chloroform) to be present everywhere he looked. In addition, he found CFC-11 to be present at almost the same concentration in all remote locations that were far removed from urban sources of pollution. These measurements strongly suggested that CFC-11 had an extraordinarily long atmospheric lifetime after its emission, was being transported over very long distances by the atmosphere, and eventually became thoroughly mixed and evenly distributed throughout the entire atmosphere of Earth.

In 1974, University of California atmospheric scientists Sherwood Roland and Mario Molina proposed a theory concerning what happens to the CFCs after being emitted into the atmosphere. They reasoned that since the CFCs were not water soluble, they would not be washed out by rain or be absorbed by the oceans and surface water. Also, they found in their laboratory that these compounds were not affected by the usual oxidation processes that break down the vast amounts of other hydrocarbons that are emitted into the atmosphere. Since nothing appeared to be removing the CFCs from the lower regions of the atmosphere (also called the troposphere), Roland and Molina suggested that they would slowly diffuse into the stagnant layer above (the stratosphere).

As the CFCs then diffused higher into the stratosphere, Rowland and Molina further hypothesized that they would eventually reach an altitude where the ultraviolet (UV) radiation from the sun would be sufficiently intense as to break their carbon-chlorine bonds, thereby forming chlorine atoms. While the original parent CFC molecules were exceedingly inert, chlorine (Cl) atoms are very reactive. Roland and Molina suggested that one of the things these Cl atoms would do was catalyze the destruction of ozone molecules which are a natural component of the stratosphere. In this way, a single Cl atom could cause the destruction of thousands of ozone molecules before it would be finally removed by some other side reaction. Because this is the only means by which the CFCs were thought to be removed from the atmosphere, the CFCs were predicted to have exceedingly long atmospheric lifetimes (on the order of a century).

Stratospheric ozone: Ozone (O_3) is composed of three oxygen atoms instead of two, as in the more common oxygen (O_2) molecule. An ozone concentration of about two ppm is continuously maintained in the central regions of the stratosphere by a set of naturally-occurring reactions driven by the incoming radiation of the sun. This layer of ozone in the stratosphere absorbs and thereby removes most of the UV portion of the sun's radiation. Without this UV screen, almost all molecules and materials on the surface of the Earth would be rapidly altered and destroyed, thus killing all forms of life that live on land. Rowland and Molina's theory predicted that a reduction in the stratospheric ozone layer caused by increasing levels of CFCs would cause a continuous increase in the intensity of UV radiation reaching the surface of the Earth. A first symptom of this effect on humans would be an increase in the occurrence of skin cancers.

Additional research between 1974 and 1985: A vast amount of additional research immediately followed the announcement of Rowland and Molina's theory; this is when I became personally involved in this aspect of atmospheric chemistry. After meeting Rowland at a conference in the summer of 1974 and learning about his theory, I immediately returned to the research laboratory I was then working in at Washington State University and applied gas chromatography/mass spectrometry, a then relatively new technique, to the analysis of the CFCs and other chlorine-containing compounds in the background atmosphere. We quickly demonstrated that we could detect and quantify all of the most important chlorine-containing components of the atmosphere. This provided the very first quantitative determinations of some of these, including CFC-12 and naturally occurring methyl chloride (Grimsrud and Rasmussen, *Atmospheric Environment*, Vol. 9, 1975). For the following few years, our lab at WSU was capable of making all of the relevant CFC measurements, and as a result, this was an exciting time for us. With the use of a Lear Jet at our local airport in Pullman, Washington, we could quickly collect air samples at all latitudes of the Northern Hemisphere and at all altitudes up to the lower stratosphere. Such measurements and many others appeared to be in line with the hypothesis of Rowland and Molina concerning the fate of CFCs in the atmosphere.

Similar and new methods of chemical detection and analysis were soon developed by many research groups throughout the world, and measurements of all related compounds were then being made in all regions of the Earth's atmosphere on a continuous basis. In addition, a more detailed understanding of all of the chemical reactions involved in ozone depletion was being obtained by their study in laboratories throughout the world. This additional research strengthened and refined the theory of stratospheric ozone depletion by the CFCs to the point where it became widely accepted by the majority of atmospheric scientists working in this area within a few years following its conception in 1974.

Public debate between 1974 and 1985: During this same period, however, there was considerable resistance within the public and industrial domains to the notion that CFCs might pose a threat to the ozone layer. The main reason for this was that the manufacturer of the CFCs, the DuPont Chemical Company, quite understandably did not wish to shut down this immense and lucrative aspect of its business. In addition, CFCs had become an important component of many industrial products and processes and these industries were also reluctant to have their supplies of CFCs terminated. In 1975, the industrial interests did agree to reduce the use of CFCs in aerosol spray cans because other methods of providing this need were readily available. In other applications, however, replacement compounds were more difficult to identify, and therefore, CFCs continued to be used in most of these for more than two decades.

DuPont and other industrial interests had little difficulty in finding scientists who were willing to state that Rowland and Molina's theory was probably not correct and that there were no environmentally sound reasons to discontinue the use and production of CFCs. The media campaign directed by DuPont between 1974 and 1985 caused a large portion of the public to suspect that the science behind the theory of stratospheric ozone depletion was seriously flawed. The chairman of the DuPont board of directors, Irving Shapiro, stated that the "ozone-depletion theory is a science fiction tale ... a load of rubbish, utter nonsense" (*Chemical Week*, July 16, 1975). As a result of this media campaign, insufficient public pressure for change was brought on the DuPont Chemical Company and it continued to be the major cause of this environmental problem — until a spectacular observation was made in 1985.

The "ozone hole" in Antarctic stratosphere first noted in 1985: In the early 1980s, reports by a British research team began to describe what appeared to be a progressively large decrease (up to 50 percent) in the total amount of ozone above the continent of Antarctica during its springtime, just as sunlight first appeared after its long, dark winter. Then in 1985, an instrument-equipped aircraft passing through the lower portions of the Antarctic stratosphere found that ozone levels were decreasing to nearly zero in vast portions of that region at that time of year. In addition, other measurements clearly showed that the cause of this ozone destruction was chlorine atoms. Further studies revealed that the precursors of these Cl atoms were produced during the prior winter on stratospheric clouds from compounds that originated from the man-made CFCs.

As the season of spring moved into summer, this "ozone hole" then spread out into regions of lower latitude, reaching New Zealand, Australia, and the tip of South America, until it gradually dissipated and disappeared for that year. (This event has now been observed every year since 1985 and is expected to re-occur every springtime for at least another fifty years. Similar but less pronounced events also occur every year in the northern polar regions.)

Replacement of CFCs: With the discovery of the Antarctic ozone hole, it became absolutely clear to almost everyone that the CFCs did in fact pose an environmental problem that simply had to be addressed. The DuPont Chemical Company then agreed to gradually phase out their production of CFCs if a global agreement to do so could be reached. By that time, DuPont was well poised for the manufacture of the various compounds that eventually replaced the CFCs in their industrial applications. DuPont's advertisements then changed to portray itself as an environmental leader in solving the stratospheric ozone problem by its development of "ozone-friendly" replacement compounds.

Slow repair of natural atmosphere: By 2002, the large-scale production of CFCs had been essentially terminated throughout the world. During the last decade, the concentrations of CFCs in the background atmosphere first stabilized (at a level of 280 parts per trillion (pptr) for CFC-11 and 535 pptr for CFC-12) and then started to decrease very slowly. Because the lifetimes of these compounds in the atmosphere are so long, it will take about a century for their concentrations to decrease to one-half their present levels. The ozone hole in the springtime of the Antarctic is not expected to disappear until about the year 2060.

What if details had been different?: We now know essentially all of the relevant details of the CFC/ozone relationship and with the advantage of 20/20 hindsight, can predict what might have happened if our response to this threat had been different. For example, if the potential destruction of ozone by the CFCs had not been recognized in 1974 and if the production of CFCs had continued to increase exponentially, then the entire world would have suffered distinctly detrimental effects by today. The intensity of UV light reaching the surface of the Earth would have risen to dangerous levels at all times of the year and the ozone holes developed at both poles of the globe would have spread across their entire respective hemispheres following each of their spring breakups. The damage thereby done to plant and animal life would have been enormous and this problem would have persisted for a couple more centuries even if we assume that the cause of this problem would have finally been recognized.

In addition, if the rates of some of the key reactions involved in these processes had just happened to be somewhat different than they turned out to be, the detrimental effects of CFCs on ozone levels could have been very much worse than they were. In retrospect, it is now clear that because of some creative thinking by a set of atmospheric scientists and some good old-fashioned luck, we Earthlings dodged a potentially fatal bullet.

Lessons learned: There are several lessons to be learned from the CFC/ozone controversy that might apply to the global warming issue being widely debated today. Some of the most obvious of these are:

- The international community of atmospheric scientists has greatly increased our understanding of the Earth's atmosphere during the last several decades, and we have no better source of information to rely on than their joint opinions as expressed through their professional organizations. In recognition of their insight that initiated our understanding of the relationship between the CFCs and stratospheric ozone, Rowland and Molina, along with Paul Crutzen of the Netherlands, were awarded the Nobel Prize in Chemistry in 1995. In addition, countless other scientists contributed to the recognition and characterization of this problem by their development of sensitive instruments that enabled us to "see" and thereby monitor all of the chemical species involved throughout our entire atmosphere.

- As in any field of science, individual scientists with contrarian views can easily be found and some of these will be strongly promoted by forces that do not wish to accept the collective views of our national scientific organizations. While clever use of these contrarian views during the CFC controversy had the intended effect of confusing the public, it is clear today that they did not represent the "best science" available at that time. These contrarian views were generally directed at the lay public, who often lacked the scientific background required for assessing the validity of what they were being told.

- The CEOs of industry chose to defend the financial welfare and investments of their employees, their stockholders, and their industry, and did so very forcefully, even though the future health and well-being of the public was being threatened by their actions. The industrial CEOs did not come down on the side of environmental and public interests until they were essentially forced to do so by a concerned public and their elected representatives.

- It is important to note that the effects of CFCs on the ozone layer became problematic even though the concentrations of these compounds only reached exceedingly low levels, 280 pptr for CFC-11 and 535 pptr for CFC-12. In arguing against the suggestion that increased levels of CO_2 will result in global warming, many have claimed that the Industrial Age increase in CO_2 (about 113 ppm) is much too small to have significantly affected the Earth's atmosphere. It should be noted, therefore, that this level of CO_2 increase is actually about 100,000 times greater than that of the CFCs. Physical conditions of our planet are indeed strongly affected by changes in the minor components of its atmosphere.

- The CFCs have atmospheric lifetimes of about one century. As we now know, some of the problems caused by these compounds will be with us for several centuries, even though their large-scale production was terminated in 2002. The CO_2 overload that we have created in the atmosphere by the combustion of fossil fuels will also have a lifetime of at least several centuries. In addition, just to feel the full warming effect of the present CO_2 overload in our atmosphere, several decades will be required due to a heat-buffering effect of the oceans. This warming effect is then expected to be with us for several millennia after we have stopped all anthropogenic emissions of CO_2. Therefore, there is an exceedingly long lag time between the causes and full effects in both of these environmental issues. The best way of predicting the magnitudes of these expected changes is by use of the best theories and models of science available, as was done during the CFC/ozone dilemma, all of which are continuously tested and refined by comparisons with both laboratory experiments and physical observations of the past.

- As illustrated by the "ozone hole" discovery, it might possibly take an exceedingly spectacular event in order to get the action needed to arrest man's effects on global warming. Apparently, the loss of most of our glaciers has not been spectacular enough. Maybe the total loss of our ice cap on the Arctic Ocean or the collapse of a portion of Greenland's ice pack along with a dramatic rise in sea levels will do the job. Unfortunately, by that time, it will very possibly be too late to solve the problem. Therefore, let's hope that corrective action will be undertaken well before then.

- As was illustrated by the worldwide ban on CFCs achieved by 1995, global cooperation can be achieved in addressing an environmental problem if it is perceived to be needed and if proper leadership is provided. The U.S. became the leader in addressing the CFC environmental problem, and it appears that problem is being solved. Even though the U.S. has been one of the major rogue countries, to date, with respect to the AGW problem, perhaps we could once again assume a position of leadership. Since the U.S.'s per-capita emissions of CO_2 are the highest in the world and because more than one-half of the CO_2 overload presently in the Earth's atmosphere came from our country, it seems appropriate that we would assume this role.

- Whenever we cause a long-term change in the composition of the atmosphere, we should try much harder than we previously have to envision all potential problems that might be caused by that change. In the example provided by the CFCs, it is now clear that two environmental disasters — not just one — would have been caused by their continued use. In addition to destroying stratospheric ozone, we now know that the CFCs would have also become major — rather than minor — greenhouse gases by today if their production rates had been allowed to increase as they were prior to 1975. If we had not leveled and then gradually eliminated the production of CFCs, we would have experienced not only enormous losses of stratospheric ozone but also significantly greater increases in temperature than we did during the last three decades. Both of these detrimental effects of the CFCs would have then been with us for a few additional centuries even if we finally did decide to discontinue their production. In short, we cannot afford to learn environmental lessons simply by the classic trial and error method. The predictive capabilities of science are essential especially when dealing with the effects of long-lived chemical substances.

Summary: Where we are today with respect to the AGW problem before us is not at all clear. Some progress has been made in making our public and elected officials more aware of this problem, but the exceedingly well-funded forces for denial and delay are still very strong and successful. We still have so many other pressing financial issues today that the issue of AGW is always in danger of being moved to a low priority. Most importantly, it is clear that sufficient actual action has not yet begun. The emissions of CO_2 throughout the world have not been reduced or even stabilized. In fact, they are still increasing. Admittedly, solving the AGW problem will be much more difficult than solving the CFC problem, because modern man has become so thoroughly addicted to a lifestyle based on fossil fuels and the peddlers of that addiction have essentially a death grip on the political system of our country. It is not yet clear that we will be able to win this one. Nevertheless, as we continue to ponder these questions, the lessons of the CFC/ozone controversy related here are well worth remembering.

CHAPTER 6

Musings of Carl Sagan on the Survival of Intelligent Forms of Life

Many of us remember the illuminating TV presentations of Carl Sagan (1934–1996), an American astrophysicist and educator, in which he introduced the public to the wonders of the cosmos. One of his primary interests was the Search for Extraterrestrial Intelligence (SETI) program by which radio receivers were (and still are) directed at various stars in our galaxy in an attempt to pick up a "call" from "someone out there." I also recall some of Sagan's speculations concerning the possibility that intelligent forms of life might exist on other planets. His thoughts went something like this:

In the entire universe, there are about 10 billion galaxies, and each of these has about 100 billion stars. This means there are about 1 billion trillion (a 1 followed by 21 zeros) stars in the entire universe. The enormity of this number — along with the fact that an intelligent form of life did develop on our planet — would lead one to suspect that the universe might be literally teeming with intelligent forms of life.

For at least a couple of reasons, however, Sagan wasn't so sure that it was. First, it takes a very long time for life-supporting conditions to develop on a favorably situated planet such as ours. Then, if an intelligent form of life finally did emerge on a given planet, Sagan thought that it might become a bit too creative for its own good and thereby cause planetary changes that would lead to its own extinction. Furthermore, Sagan's observations of life on this planet made it clear to him that intelligent beings could manage to self-destruct in an exceedingly short time, essentially in an instant on the geologic time scale. When he included this latter factor in his models of the universe, the probability of someone being "out there" plummeted.

While it is not difficult to imagine a number of ways by which the intelligent forms of life that presently exist on this planet could do themselves in, Sagan was particularly concerned about the potential problem of AGW long before it became the major issue it is today. His own studies of our nearest neighbor, Venus, during the 1960s, provided him with a spectacular example of greenhouse gas warming by carbon dioxide (CO_2). The surface temperature of Venus is known to be a sizzling 900 degrees F and its atmosphere contains about 96 percent CO_2. The existence of river valleys is also evidence that water was on the surface of Venus at some point, and this suggests that it once had more moderate Earth-like temperatures.

Sagan speculated that at some point in its development, Venus must have undergone a "runaway" greenhouse effect. That is, as the planet began to warm for some initial and unknown reason, increased amounts of CO_2 were emitted from the surfaces of Venus. This additional CO_2 then caused additional warming, thereby causing the emission of more CO_2, causing more warming, and so on, until the process finally stabilized at its present inhospitable state from which it can never recover. Fortunately, scientists believe that the initial conditions that led to the extraordinarily high temperatures reached on Venus do not yet exist on our planet. Nevertheless, Sagan also recognized that the increased levels of CO_2 that are expected to follow the combustion of the fossil fuels on Earth would very possibly lead to temperature increases that would make our planet unsuitable for its human inhabitants.

As to why Sagan thought this might occur in a geologic instant of time, one must consider the issue from the long-term perspective. The Earth is thought to have started forming about 4.5 billion years ago. After about 4 billion years of cooling and the slow development of simple microbial forms of life, the atmosphere finally became more conducive to the evolution of plant life by the accumulation of some CO_2, possibly provided by volcanoes. By way of photosynthesis, more complex forms of plant and then animal life began to appear about 500 million years ago. After death, some of that plant material became progressively covered with dirt and water, thereby preventing its immediate decay and conversion back to the atmosphere as gaseous CO_2. Instead, continuously increasing pressure and geothermal heating of that matter caused its slow conversion to the fossilized forms of carbon we are familiar with today, including gas, oil, and coal.

The speed with which man has already converted much of these fossilized forms of carbon back into atmospheric CO_2 is truly astounding. Significant use of fossil fuels by man began only about 160 years ago, at the beginning of the modern Industrial Age. From then up to the present time, we think that we have consumed about half of the Earth's reservoirs of oil and a significant fraction of its coal. If consumed at the present rates, it has been estimated that our reserves of coal will last less than two more centuries, and those of oil only several decades.

Therefore, while it took about 300 million years to convert massive amounts of atmospheric CO_2 into the fossilized forms of carbon, man has shown that he has the ability and possibly the inclination to convert a very large fraction of that back to atmospheric CO_2 in just 300 years! If that enormous conversion of carbon from its fossilized to biological forms is allowed to occur, the resulting effect on the Earth's energy balance will be unlike anything ever experienced during the history of man on this planet. The increased temperature caused by this enormous overload of CO_2 would very likely be sufficient to put the Earth into its own version of a runaway greenhouse effect. In the Earth version, not only CO_2 but also our other two major greenhouse gases, methane and water vapor, would increasingly be emitted into the atmosphere from their abundant surface reservoirs as the temperature rose. While the Earth version is not yet expected to go all the way to the extreme point that Venus's runaway effect did, ours could bring us back to its "water world" condition of 35 million years ago when there was no ice on Earth and sea levels were about 70 meters higher than today. Moving in that direction relatively quickly on the geological time scale would, of course, be exceedingly detrimental to existing forms of civilization.

There is no question that the intelligent form of life that developed on this planet was sufficiently intelligent to create a good standard of living for a large fraction of its members and a lavish standard for many. The critical question remaining today, however, is whether Earthlings will prove to be sufficiently intelligent to prevent the demise of their own species during the upcoming decades. During his lifetime, Carl Sagan did not find evidence of intelligent life existing elsewhere in our galaxy, and based on his observations of life on this planet, seemed to understand why he might not.

CHAPTER 7

Churchillian Leadership Needed for the Next Decade

"These are the times in which a genius would wish to live. It is not in the still calm of life that great characters are formed," Abigail Adams, speaking of another critical moment in American history.

In the first decade of the 21st Century, almost no efforts were made by the leaders of our country to address AGW and since the last presidential election of 2008 only small improvements have been made. During the first three years of the Obama administration, his mediocre record of accomplishment in this area can be partially attributed to the myriad problems that he faced when he took office. First, our faltering economy and the threat of a 1930's-like depression seemed to occupy most of his attention. We still had two military confrontations to deal with in Iraq and Afghanistan and he has had to remain alert to potentially destabilizing events occurring elsewhere in the Middle East. Because the high costs of our wars in the Middle East have been put on our national credit card, we have yet to figure out how we are going to pay for them in the future along with our existing social security and health care programs which, by themselves, are expected to consume the entire national budget within two or three decades. The long-neglected physical infrastructure of our country is also in dire need of major repairs and upgrades. In addition, we seem to be continuously dissatisfied with many aspects of our educational systems which may also need large additional investments of human and financial resources. And finally, President Obama has directed a lot of his time and energy towards the health care system of our country.

Therefore, what is certain to become the greatest and most challenging problem facing us in the 21st Century, AGW, still awaits us and, in fact, has become even more formidable during the Obama administration. Most people wish that this "800-lb gorilla in the back of the room" would simply go away until our other immediate problems can be addressed. Unfortunately, the profound nature of the AGW problem is such that if not adequately addressed in a very timely manner, all other endeavors of mankind are likely to be rendered inconsequential within the next few decades. Therefore, the problem of AGW must immediately be given the highest priority within our nation and the entire world.

In order to stand a chance of success in this fight, it is now clear that at least two things must happen. One is that essentially all global emissions of carbon dioxide by the combustion of fossil fuels must be reduced immediately and then stopped entirely by the end of this century. The other is that extraordinary levels of national and international leadership will have to emerge. In a recent address in my community, climate change expert Dr. Steven Running of the University of Montana stated that he believed that the first of these two requirements will be met — that is, technical solutions for replacing carbon-based energy will be found. However, he was not at all optimistic that the second requirement will be met. That is, he doubted that sufficient leadership would emerge soon enough to sufficiently cut carbon dioxide emissions.

It is easy to understand why the required levels of leadership might not emerge — one needs only to envision some of the specific tasks that such leaders would have to undertake. For starters, they might have to persuade the public to accept higher prices for all use of fossil fuels so that alternative sources of energy would become financially more attractive. These leaders might have to convince us to make significant changes in our lifestyles. For example, they might have to ask us to reduce or eliminate much of our unnecessary travel by fossil-fuel-powered vehicles. They might have to insist that we not use our nation's abundant supplies of coal for energy or for liquid fuel production (unless the CO_2 thereby produced can be captured and permanently sequestered). The reason for this, these imagined leaders would have to explain, is because ironically there is simply way too much coal remaining on Earth for its atmosphere to handle. These leaders would also have to prevent the development of other new and abundant sources of fossil fuels, such as the tar sands of Alberta, again simply because our atmosphere cannot stand the additional carbon dioxide that its use would cause. They might even have to ask us to not try so hard to develop new sources of gas and oil, the relatively clean and high-energy forms of fossil fuel. They might also have to ask us to cut back our per capita carbon emissions much more than the citizens of essentially all other countries because our past and present per capita emissions are two to several factors greater than theirs. These leaders would have to convince the CEOs of industry to make drastic changes in their current modes of operation which have been developed over many decades of fossil fuel use. While these changes would lead to the development of new industries, they might also lead to the demise of others that are presently very powerful and exert tremendous influence over public policy. While convincing the U.S.A. of the necessity of these measures, these leaders would also have to convince other countries to do the same. In short, at least one of these envisioned leaders would have to be a world statesman of the very first order, possessing degrees of wisdom, dedication, determination and credibility that are rarely seen in a given generation.

In trying to envision such leaders, it is instructive to reflect on outstanding leaders of the past who might serve well as models for our needs today. The examples of Abraham Lincoln, Franklin Roosevelt, and Winston Churchill immediately come to my mind. All of these leaders were willing to focus primarily and forcefully on the single greatest issues of their times: Lincoln is known for his insistence that the Union not be divided, Roosevelt for forcefully addressing the Great Depression, and Churchill for recognizing the threat posed to western civilization by Nazi Germany. In order to face their distinctively up-hill battles, all of these leaders had to expend major portions of their political capital and place themselves squarely in the "bull's eye" of their equally determined opponents. All of them became heroic figures in the public eye only after their missions were either well underway or accomplished. I have chosen to focus on Churchill's example below because of the very long period of almost total public resistance and denial he faced. Another reason for choosing the Churchill example is that it clearly illustrates how denial and a lack of timely action can greatly magnify the size of a potential problem. That is, WWII did indeed turn out to be a very big deal in the 40's — while it could have been prevented entirely by appropriate and far more moderate actions in the 30's.

In 1918, Britain had just emerged from WWI in a distinctly somber state. A large fraction of its male population of military age had been either killed or wounded during that gruesome four-year war, and the material cost of that conflict had severely depleted British reserves. In the prosperous decade of the 1920s, Britain had just begun to rebuild itself and its long-standing British Empire on which "the sun never set." Then the financial crash of 1929 sent Britain and the entire Western world into a deep depression that was to last throughout the 1930s. As Britain struggled with these economic problems, momentous changes were occurring in Germany.

Most significantly, Adolf Hitler and his Nazi party took control of Germany in 1933. He embarked on an ambitious re-armament program, invoked anti-Semitic laws, retook the Rhineland in 1936, formed the Nazi Axis alliance with Italy's Mussolini, and then seized the neighboring countries of Austria and Czechoslovakia in 1938. During this entire period, His Majesty's Government (HMG) of Britain followed a policy of disarmament and appeasement in order to not upset Hitler, thereby avoiding another war with Germany.

Starting in about 1932, however, another small but significantly influential group of British citizens, led by Winston Churchill, had a very different view of events in Germany. They saw the threat posed by Hitler's Nazis for what it later proved to be, and did everything in their power to bring the rest of their country to their point of view. They had little success in this endeavor, however, throughout most of that decade. In fact, even after Hitler's annexation of Austria and Czechoslovakia in 1938, the policy of appeasement ruled and Churchill was vilified roundly for his efforts, commonly labeled a "fear monger" who was irresponsibly frightening the public with his message of impending doom. Some of Churchill's political adversaries declared him to be the greatest threat to peace and prosperity in the world. Also the majority of the British people could not yet bear the thought of undertaking another war with Germany and preferred to follow the more comforting path of denying and ignoring the menace posed by their Nazi neighbor.

Only after Germany invaded Poland in 1939 did HMG finally acknowledge that Churchill had been correct concerning the Nazi threat and finally declared war on Germany. As Germany then easily rolled over the ill-prepared forces of France and Britain on their western front in the spring of 1940, HMG finally asked Churchill to take over the reins of government and try to save their island nation from Germany's further advance.

It is only after this point in history that the reputation of Winston Churchill was restored. It then soared as he led Britain through its "finest hour" that slowly began to turn the tide against Nazi expansion. During this period, Churchill honestly related to the British people what would be required of them. He offered them only "blood, toil, tears, and sweat" in order to bring their country through this initial dark period of WWII. His basic message was "we must never surrender" in this battle because "without victory, there is no survival."

The citizens of Britain then responded. After suffering the hardships Churchill foresaw, Britain just barely managed to halt Germany's westward advance. By her example, Britain prompted other Western nations (and most importantly, the U.S.) to subsequently enter the global war that eventually led to the destruction of Nazi Germany in 1945.

By following the advice of Churchill, Britain did indeed suffer many of the detrimental consequences that the British had feared might accompany another war. They lost not only many military personnel, but also civilians and portions of their major cities. Their dwindling material resources were further diminished and the disintegration of the British Empire then proceeded much more rapidly. After WWII, as a result of these setbacks, Britain was no longer one of the most powerful countries in the world. The British people's reward for these sacrifices was the survival of their free and democratic way of life.

All of this occurred, one could argue, because of Churchill's sense of priorities and his monumental determination. Most historians agree that Britain and the entire Western world were extremely fortunate to have had this leader in their midst during this critical period of the twentieth century. It seems to me, that leaders of Churchill's caliber are again desperately needed today. That is, leaders who understand the paramount importance of the survival of the civilized world and have the wisdom and courage required to set their priorities accordingly.

CHAPTER 8

Getting Past the Forces of Denial

At this point, all of the leading scientific organizations of the U.S. and the Intergovernmental Panel on Climate Change (IPCC) of the United Nations have stated unequivocally that AGW is now occurring and that it should be addressed in a timely manner. Nevertheless, this message has not yet been effectively transferred to the American public. A significantly large portion of the American public and their elected officials still tend to deny, ignore, or downplay the importance of anthropogenic global warming (AGW). While the policies used to address complex environmental and technical problems should ideally be based on the very best science available, in a democracy such as ours, the endorsement and implementation of those policies ultimately lies with the voting public and their elected representatives. Therefore, in this chapter, I will attempt to identify and discuss some of the reasons why it is so difficult to accept the messages of our scientific community concerning AGW. Many of these observations have resulted from personal conversations I have had with hundreds of people.

First, the big one: "Not if my ox will be gored." One of the inevitable consequences of the technological changes that have occurred throughout the history of civilization has always been that they might disrupt one's means of livelihood. It is very difficult indeed to convince either a single individual or a large industry that something must be done if doing that thing could result in the elimination of that person's job or that entire industry. As painful as this process always is, however, it is one that usually must be accommodated in order to develop better technologies that address substantial problems as they emerge.

In contemplating this dilemma, I am reminded of my grandfather's brother, Lucas Grimsrud, who was a young blacksmith in Coon Valley, Wisconsin. In 1900, he decided to "go west" in the hopes of finding increased opportunities in his trade. Upon arriving in Spokane, Washington, he set up a blacksmith shop in what is now its commercial center. I can imagine that he was not thrilled when the mass production of automobiles then grew rapidly in the U.S. in the following years. While Lucas appeared to have adjusted over the subsequent years by gradually changing his Spokane business to auto body repairs, I suspect that many blacksmiths of that era did not do as well.

While new opportunities are created by change, the financial victims of that change are not necessarily well poised to take advantage of them and, instead, will often use their energy and influence to resist that change. Many of the CEOs of the fossil fuel industries have provided us with classic examples of this. These individuals were sufficiently intelligent and sufficiently well informed as to have realized a long time ago that the increased use of their products would very possibly lead to a major environmental problem for all citizens of the world. Nevertheless, the CEOs who remained in power did not generally acknowledge the fundamental importance of this problem until very recently, after intense media coverage began in about 2006. Since then their strategy seems to be to admit that AGW constitutes a problem, but also to delay any decreases in fossil fuel use. These facts call into question whether or not the CEOs of large industries are capable of using their social consciences when doing the "right thing" might not be financially beneficial to their industry and stockholders and, therefore, might lead to their own dismissal. History suggests that large industries squarely address environmental problems only when they are forced to do so by the application of sufficient pressure from the public and their elected representatives.

"Don't be concerned, we are taking care of it": Only a few years ago, the fossil fuel industries and their associated think tanks were endeavoring to undermine the science behind AGW. Most of them have now changed their approach by acknowledging in their more recent advertisements the need for alternative sources of energy. In addition, these advertisements give the public the impression that the fossil fuel industries are working very hard on the problem and that everything is now under control. The implication left with many is that we should no longer worry about this issue and, in the meantime, should continue to buy, consume, and travel as before. One particularly misleading example of this new advertisement mode is the assurance by the coal-fired power industry that in the future they will be utilizing "clean coal" technology.

Advertisements such as these serve their intended purpose of soothing the public and making them less inclined to press for substantial changes in energy production. Now consider the fact that only a tiny fraction of the total resources of the American fossil fuel industries are presently being devoted to alternate sources of energy.

Also consider the fact that "clean coal" technology (that is, non-CO_2-emitting power plants) does not yet exist as a financially viable alternate — and very possibly never will. In light of this, perhaps one should not be at all comforted by this new form of advertisement used by the fossil fuel industries.

"Let the free-market system handle it": A seemingly attractive approach to the AGW problem has been suggested by Bjorn Lomborg in his two books, *The Skeptical Environmentalist* and *Cool It!* He acknowledges the occurrence of AGW and recommends the development of alternate energy sources. However, he also argues that we should continue to use our fossil fuels until the alternate sources of energy become as financially attractive as the fossil fuels. At that point in time, the entire world could switch to these non-polluting energy sources without incurring a financial penalty in the meantime.

The viability of this approach depends, of course, on how long it would take for this financially driven switch to occur. The day when the alternate forms of energy will become as cheap as coal is clearly a long way off in the future, and even just ten more years of business as usual is environmentally unacceptable. With the advantage of hind-sight, it is now clear that we should have taken up Lomborg's suggestions more than 50 years ago. It is now much too late to do so, however. Far stronger and more immediate actions are now required.

Is there a "malaise on the land?" During a previous energy crisis in 1979, President Jimmy Carter gave a nationally-televised address in which he identified what he believed to be a "crisis of confidence" among the American people. This became known as his "malaise" speech, although Carter himself did not use that word. In his speech he said: "In a nation that was proud of hard work, strong families, close-knit communities, and our faith in God, too many of us now tend to worship self-indulgence and consumption. Human identity is no longer defined by what one does, but by what one owns. But we've discovered that owning things and consuming things does not satisfy our longing for meaning."

I will add my own two cents to Carter's thoughts concerning what might be happening to some today. Many people seem to view all environmental problems as a part of social decline, in general, driven by a rising tide of selfishness, greed and corruption. They feel powerless to do anything about it so their concern translates into frustration rather than support for action. They find solace primarily in their private domain and cease to consider themselves a part of their larger state and national communities.

What we are dealing with here might be an impression of reality more than reality itself. This might help explain the changes in attitude that some felt shortly after 1980, when Ronald Reagan became President. No one will dispute the fact that one of President Reagan's greatest virtues was his unquenchable optimism concerning his country's ability to face hardships. We should perhaps be pleased that one of our current president's models for effective leadership is Ronald Reagan.

Media coverage of climate change: It is not difficult to understand why the issue of AGW was not taken more seriously than it was by the majority of Americans prior to about 2006. Until then, this topic did not enjoy a sustained level of media attention and the public is, indeed, greatly affected by the news they are daily exposed to.

During the Reagan era (1981–1988), some of the first alarms

were sounded concerning the likelihood of rapid climate changes. These came primarily from scientists who were developing detailed models of Earth's atmosphere made possible for the first time by rapidly improving computer technology. At that time, however, the primary focus of President Reagan and the media was on the cold war that soon led to the collapse of the Soviet Union. Finally, in 1988, there was a flurry of media attention concerning AGW created by the testimony of Dr. James Hansen, head of the NASA Goddard Space Science Research Center, before the U.S. Senate. Dr. Hansen reported that their computer-based models indicated that global warming was occurring much more rapidly than had been previously thought and that a large portion of this warming trend was due to the impacts of man on the atmosphere. This initial high level of media coverage only lasted a couple of years, however.

During the subsequent twelve years of the George H.W. Bush and Bill Clinton administrations (1989–2000), relatively little leadership on the subject of climate change was provided by either of these presidents. Media coverage of climate change was also relatively low at that time possibly because the visible effects of global warming had not yet become as apparent as they were to become in the first decade of the twenty-first century. It is interesting to note that when Al Gore ran for president in 2000, he chose not to emphasize his concern about AGW. He and his staff apparently decided that this issue was too complex and still too theoretical at that point for the voting public to accept. Similarly in his presidential campaign of 2004, John Kerry placed only a moderate level of emphasis on AGW. In addition, the president from 2001 to 2008, George W. Bush, did very little to raise public awareness of the global warming problem. In fact, he endeavored to muzzle or tone down messages of concern on this topic coming from senior scientists within his own government agencies, including the EPA and NASA.

For several reasons, this issue has again received a great deal of media attention since about 2006. First, the United Nations' Intergovernmental Panel on Climate Change (IPCC) had been turning out increasingly grim reports every three years concerning the AGW problem, and their report in 2007 was particularly alarming. Al Gore's movie, *An Inconvenient Truth*, also did much to bring the AGW issue before the American public. The subsequent 2006 Nobel Peace Prize awarded jointly to the IPCC committee and Al Gore provided additional attention to this message. Perhaps most importantly, in the first decade of the twenty-first century, additional physical evidence, particularly in the polar regions, made it increasingly clear that an unprecedented level of rapid warming was occurring on our planet.

It should not be assumed that the intense level of media coverage of the last few years will continue, however. The great economic problems that steadily worsened throughout 2008 still threaten to push the issue of AGW back to a level of lower priority than it deserves and requires. Hopefully, that trend will be reversed after the U.S. Presidential election of 2012.

Political affiliations: Many people tend to adopt positions on various issues that are endorsed by their favored political party. Since the presidency of Ronald Reagan, there has been a tendency for the Republicans to be less favorable to environmental issues than the Democrats. It is instructive to note that this difference was not apparent during the first portion of the twentieth century. For example, at the very beginning of that century, the Republican Theodore Roosevelt was the first president to demonstrate concern for the long-term conservation of our natural resources. Roosevelt was instrumental in forming the National Forest Service and our National Park System. It should also be noted that the Environmental Protection Agency was suggested and created by President Nixon in 1970 and that the Conservative Prime Minister of Great Britain, Margaret Thatcher, defined global warming to be an issue of key importance within her country in 1988. Whether one considers himself to be a political liberal, conservative, or moderate, it is difficult to understand why that choice should make a difference with respect to one's hope that the natural environment remains healthy and intact for future generations. After all, shouldn't a conservative be just as interested in *conservation* as anyone? That is, doesn't the long term financial value of our collective holdings on this planet require our constant care and maintenance?

Relative wealth of individuals: A great difference in the distribution of personal wealth has developed in the U.S., especially over the last few decades. It is useful, therefore, to consider whether the factor of relative wealth has an influence on the inclination of one to accept or reject the notion of AGW.

There is no doubt that the wealthy among us tend to have more lavish lifestyles than those who can only afford relatively modest ones (as measured by the usual parameters such as number and sizes of homes, number and type of cars, RVs, and boats, amount of travel, etc.). It seems likely, therefore, that in order to address the causes of AGW, the wealthy will be encouraged to make greater adjustments to their lifestyles than those who already have modest lifestyles. On the other hand, the relatively poor will not be as able as the wealthy to make necessary improvements in the efficiency of their energy-consuming devices and homes. Therefore, both of these groups will be challenged to make their own contributions to a total reduction in energy use.

It is interesting to note that these financial differences among us are not necessarily related to political affiliations. There are both conservatives and liberals who enjoy lavish lifestyles, just as there are both conservatives and liberals who live modestly.

Many of us will undoubtedly continue to feel that we are entitled to enjoy the fruits of our labor as we wish, and in our country, this has certainly been true. Nevertheless, if we are to seriously address AGW, it will become necessary for all of us to learn more environmentally friendly ways to use our personal resources.

Man's unwarranted hubris concerning his historical experience: Another common reason for denial of the threat posed by AGW seems to be related to man's gross overestimation of his own historical experience on this planet. One often hears, for example, expressions such as, "We've been through some tougher times before," "We will surely meet this challenge just as we have before," or "I have heard these doomsday scenarios before and they never amounted to anything." Statements such as these give the impression that man thinks he has been around a long time and has gained a great deal of relevant experience and wisdom. He tends to think that he is prepared to solve just about any new problem he might be faced with and certainly is not going to be "fooled" by yet another cockeyed prediction of impending doom.

The great flaw in this line of thinking is that the total sum of all human experience and knowledge has been acquired only over an exceedingly short period of geologic time. Whether one views the period of man's education concerning his environment to extend back 100 years (to the beginning of the twentieth century), 240 years (to the founding of our country), 2,000 years (to the Roman Empire), or 10,000 years (to the very beginning of human civilization), the length of all of those periods constitutes little more than a tick of the clock on a geological time scale. The lessons learned during these exceedingly short periods of time have definitely not prepared us for the global warming problem we face today.

No comparable changes in Earth's composition of greenhouse gases and the resulting energy balance have ever occurred before in the history of man's experiences on this planet. Nor have such changes as these occurred in the roughly three million years of glacial / interglacial oscillations prior to the Industrial Age of man. Therefore, it is important that man realizes he has never faced a problem of this magnitude before and it is unlikely that he will find its solution in the "bag of tricks" he has acquired from the past.

Dealing with the guilt: One of the dilemmas associated with accepting the notion of AGW is that in doing so, we might also be admitting that our past, present, and even future behavior is contributing to the degradation of our planet — and possibly to a compromised future for our descendants. This can be a difficult admission to make and it is, indeed, tempting to take an easy way out. One means of escape is to adopt an intentional form of ignorance and simply claim that we did not know about the dangers of AGW. We can always say, for example, that the case for AGW was never presented to us in a sufficiently clear and/or convincing manner and, in addition, we thought "even the scientists were divided" on the issue.

I think a much better approach to this dilemma, however, is to realize that the condition we now find ourselves in was long and gradual in the making and, quite understandably, very few of us realized what was happening. After World War II, an American culture was progressively developed that depended on continuous growth and increased consumption. This cultural trend was viewed as an admirable American quality by most of the baby boomers who were raised during this period. While life was indeed good in a materialistic sense, we lacked the collective wisdom needed to understand the long-term consequences of this trend. Now our perception of our past and future is changing, however, and we should simply focus on doing what needs to be done from this point on.

Recognizing the urgency: Most of us have a hard time thinking about issues that have long time scales, such as several decades to centuries. Hence, many of us tend to ignore or downplay the importance of AGW because we don't sense that it will affect our day-to-day lives right now or even next year. Also, scientists are not able to tell us exactly when and how climate changes will affect the locations where we happen to live.

Nevertheless, I am sure that we all care a great deal about some long-term issues in our lives, and perhaps the most important of these is the continued welfare of our children and grandchildren. Is there a more important investment we can make to ensure bright futures for them than to maintain favorable conditions on their planet? Even though we don't know exactly when the catastrophic effects of unaddressed AGW will occur, there is little doubt that they will occur in this century and will affect the youngsters we are just now getting to know and prepare for life.

Overcoming our nationalistic instincts: Historically, the U.S. has tended to resist international efforts to unite the countries of the world in order to address common interests. Woodrow Wilson's efforts to form the League of Nations came to naught when the

U.S. Senate failed to ratify the agreement he had orchestrated in Europe after World War I. The United Nations was finally formed immediately after World War II, but it has often not been held in high regard within the U.S. Conspiracy theories concerning unspecified international forces trying to take control of the world are constantly in circulation today. As James Lovelock suggests in *The Revenge of Gaia*, a basic sense of tribalism and "us against them" seems to be in our genes.

Nevertheless, AGW is the sort of problem that can be solved only by international cooperation, global consensus, and coordinated action. Our planet is no longer sufficiently large and robust as to allow us each to do whatever we please. For this reason and undoubtedly for many others, we must view the citizens of all countries to be our partners in this joint venture to preserve favorable conditions on our planet. In short, we must learn to view our global community as being one of primary importance.

"But we don't want to give up more of our person freedoms": Yes, one does often hear that we will suffer additional losses of our personal freedoms if we let our government play a central role in energy policy. Upon closer inspection, however, one soon realizes that is not really true. On issues as huge as that of energy policy, a great deal of central coordination and control by someone is absolutely required. The only question is, who are you going to give that central responsibility and power to? Other than our elected government, the only alternative is Corporate America who will, indeed, gladly continue to serve that function for us. Because the foremost goal of Corporate America is to serve the immediate financial interests of their stockholders, however, and is only weakly linked to the longer term interests of the general public and future generations, you already know what type of coordination and control they will provide — the same as they have provided in the past. I believe that we have already turned far too many of our personal freedoms over to the business-as-usual forces of Corporate America and must put our own government in charge of our national energy policies.

"Leave it to God": For some people, I have noted that aspects of their religious beliefs or personal philosophies prevent them from entertaining the possibility that man has the power to affect the future of our planet. Some also seem to believe that whatever happens is simply God's will. I have little to contribute to this discussion other than to offer my own expectation that any benevolent deity would prefer that we be good stewards of the gifts He has given us and that we do our best to preserve these gifts for future generations.

In considering this topic, I am reminded of an example again provided by one of my grandfather's close relatives — a cousin named Herman Ekern. In 1911 Ekern became the Insurance Commissioner for the State of Wisconsin and, due to his own background, was particularly concerned about the financial well being of the rural population of the Midwest. Therefore, in 1917 Herman and his counterpart in Minnesota, Jacob Preus, gave birth to a new insurance company initially called Luther Union but renamed three years later to Lutheran Brotherhood (it is now called Thrivent Financial). The major obstacle Ekern and Preus faced in getting this organization started was to overcome the view commonly held by its Norwegian participants that it was sacrilegious not to leave one's fate in the hands of the Lord. They were successful in getting this organization started, however, and eventually it became one of the largest fraternal benefit societies in the United States. Evidently, the notion that *God helps those who at least try to help themselves* eventually did win out among the Norwegian Lutherans of America. With respect to the AGW issue, hopefully all among us who believe that *God's will indeed be done* will also come to that point of view.

"But it's already too late": If one reads, understands, and ponders the scientific literature concerning AGW, one is struck by the real possibility that it might already be too late to reverse the warming trend that we have already begun. After all, we know that the rate of GHG emissions is still increasing beyond the unacceptably high levels of the past. On top of that, we also know that even if we could stop all GHG emissions tomorrow, the full warming effect of the GHGs that have accumulated up to today will continue to play out for many centuries. As a result, there are undoubtedly numerous people who think our planet is already out of control with respect to its continued and then irreversible warming. These individuals probably tend to keep their opinions either to themselves or within a small group of companions. What would be the point, after all, of crying "fire!" in a confined building if there was no means of extinguishing the fire — and no means of exiting the building?

Nevertheless, one of the world's most experienced atmospheric scientists, Dr. James Lovelock of Britain, has come to a distinctively grim view of the world's future and is now openly expressing it. Dr. Lovelock is perhaps best known for his "Gaia" hypothesis concerning Earth's continuous attempt to heal itself against the injuries inflicted by man. He has published more than 200 articles and books in the field of environmental science, and in the early 1970s, reported the first detections of CFCs in the background atmosphere. Dr. Lovelock is still going strong at 93 years of age, and his opinion is definitely one to be taken seriously.

In his most recent book, *The Vanishing Face of Gaia*, Dr. Lovelock states his belief that humans are simply incapable of reacting fast enough using the technological approaches to energy production that are being considered today and therefore will not be able to prevent enormous changes in climate during this century. Furthermore, he estimates that by the year 2100, the world will be able to support no more than about 1 billion people and that these survivors will be living at much more northern and southern latitudes than we presently do, because the mid-latitudes will have been reduced to deserts.

According to the World Population Balance, an advocacy organization in Minneapolis, today's population of almost 7 billion is presently increasing at a rate of 200,000 per day and is expected to reach 9 billion by 2050. Thus, if Dr. Lovelock's prediction of less than 1 billion inhabitants by 2100 turns out to be correct, there will likely be an unimaginable level of human suffering and societal breakdown awaiting us in the latter portion of the current century.

This dim view of the future has also now penetrated into the public's common forms of communication and entertainment. For example, on his show of September 9, 2008, the late-night talk show host David Letterman sincerely, if also darkly humorously, expressed his own strong opinion by ranting, "We've had no leadership in this issue since it began in the 1980s," and now "It's just too late, folks, we're all dead meat!"

While I agree that the probability for occurrence of this grim outcome is a number significantly greater than zero, my own understanding is that we still have a reasonably good chance of arresting and minimizing the warming trend caused by man if immediate and forceful actions are taken. The old saying, "all we can do is try" takes on new meaning indeed when applied to the problem of AGW.

CHAPTER 9

So What's Ahead for Us?

"Hope isn't the kind of thing that you can say either exists or doesn't exist. It's like a path across the land that's not there to begin with, but when lots of people go the same way, it comes into being," Lu Xun, Chinese writer

I will not attempt to describe in any detail here what is likely to happen on our planet if we fail to adequately address the threat of AGW. For those who want to learn more about the details of these predicted outcomes, I recommend Mark Lynas's book, *Six Degrees: Our Future on a Hotter Planet*, which is based on the most recent literature of climate change. From my own inspection of that literature, I will share just one thought here: I am very concerned, indeed, about the well-being of future generations, beginning with that of my own five grandchildren, Charlie, Kate, Emma, Elsa, and Krista, ages 4 years to 4 months. If we do not forcefully begin to address the causes of AGW **during the next decade**, it is a virtual certainty that by the time my grandchildren have graduated from high school, the concentration of carbon dioxide in our atmosphere will be at least 450 ppm, a level not seen in 35 million years. By the time they reach their 30's, the

Earth's climate system might already have entered an unstoppable autopilot mode in which it moves irreversibly towards more warming. The remaining portions of their adult lives would then, of course, be exceedingly problematic. Perhaps the worst aspect of the dilemma they would then face would be the realization that their problems would no longer be solvable by any imagined means no matter how forcefully applied. Of course, no one wants this hopeless condition to be the overriding legacy we leave to our grandchildren and their descendants. It is absolutely essential to realize, therefore, that the only opportunity humanity might have to prevent that presently unimaginable outcome is very likely to be **in the next decade.**

It is also beyond the scope of this book to describe and assess in detail the array of technological options that will hopefully help us prevent dire outcomes such as those summarized above. I will instead provide a general overview of four major areas that I believe are most likely to provide the changes required. The first two of these concern the developments of two major alternate sources of energy: those derived from the sun and those derived from nuclear fuels. The third factor concerns the issue of increased efficiency of energy use. The fourth factor concerns which of our fossil fuels we should use during our transition to a fossil-fuel-free era.

Energy from the sun: Man has always tried to use the power of the sun to provide at least a portion of his energy needs and with the great technological advances that have been made in the last century, this trend will certainly continue. The energy derived from some of these innovations, such as photovoltaic and solar thermal panels, obviously comes directly from the sun and that derived from others, such as windmills and bio-fuels, does also if in a less obvious manner. Because the amount of energy being delivered to the Earth by sunlight is about 120,000 terawatts at all times and the total energy needs of all human beings is currently about 16 terawatts, we are receiving an ample amount of energy from the sun, alone, to provide our needs if we continue to learn how to use it effectively.

Future dependence on solar energy is also advisable from another standpoint, not yet widely appreciated. All energy use ultimately results in heat being dissipated to our planet. If that energy is derived from the sun, no net heating of the Earth will thereby occur, however. We will simply be using that energy for performing useful work prior to its dissipation into heat. If, on the other hand, the energy we use is derived from any source other than the sun, such as from the combustion of fossil fuels or from the nuclear reactions of elements long ago deposited on the Earth, the heat thereby generated is extra heat over and above that provided by the sun. A term for this extra amount of heat delivered by non-solar power sources is "waste heat" which will be additionally described in the following section dealing with nuclear power.

Thus, this brief section concerning future uses of solar power can be easily summarized with the words, "go for it!" While there will undoubtedly also be some downsides to these approaches, they should be manageable.

Energy from nuclear reactions: With the advantage of hindsight, it is now clear that the United States should not have halted its development of nuclear power plants some 30 years ago. This did not occur, by the way, merely because of the objections of Jane Fonda or any environmental movements at that time. It occurred because the power companies correctly saw that power generation by nuclear reactions could not compete with that generated by the combustion of fossil fuels at that time. This is still true today largely because the nuclear industry must include the costs of the disposal of their waste products while the fossil fuel industry does not. As will be described in Chapter 10, this free service to the fossil fuel industries must be corrected so that a level playing field is created for the future development of both nuclear and solar-based sources of energy.

In describing the historic development of nuclear reactors for power generation, it is useful to divide the plants thereby built or envisioned into four classes. Class I plants were based on relatively primitive graphite-pile reaction cores first developed during the Second World War for the purpose of making the fissionable nuclide, Plutonium-239, that was used in two of the three atomic bombs first developed in 1945. The nuclear reactor that exploded at Chernobyl in the Russian Ukraine in 1986 was a Class I reactor. The Class II reactors were also developed in the mid-twentieth century and include most of those in use today. They are based on much safer and more controllable water-based reaction cores. The reactors that were recently overwhelmed by the tsunami that hit Fukushima, Japan, in 2011 were 30 to 40-year-old Class II reactors. The very few new reactors that are being built today are Class III reactors. They are based on the same principles as Class II reactors but include new technologies for increased safety, control and useful lifetimes. Class IV reactors are still in their development stage and various versions of them are expected to come on board gradually in the next several decades. They are generally based on the use of a liquid metal (such as sodium at high temperature) as the core moderator. These "fast-flux" reactors, as they are called, are capable of "burning" all of the heavy elements put into their core. For example, naturally-occurring uranium contains less than 1% of its isotope 235 and this isotope, only, undergoes nuclear fission in conventional reactors of the types I, II, and III. In a class IV reactor, uranium-238 which constitutes over 99% of natural uranium is converted to other heavy nuclides which do then also undergo nuclear fission. In this way, all of the uranium added to a Class IV reactor is burned. Note also that the Earth contains an almost inexhaustible supply of uranium-238 and other heavy elements that can be used in Class IV reactors. Even the massive amounts of waste products accumulated from our prior use of the older reactors can be used as fuel for Class IV reactors. Another great advantage of Class IV reactors is that their radioactive wastes are much easier to deal with. They produce far fewer radioactive wastes and these consist only of elements in the middle of the periodic table with radioactive half-lives much shorter than the waste products of the conventional reactors.

While I believe it was an enormous mistake to discontinue our research and development in the area of nuclear power plants some 30 or more years ago, we did not know then as well as we do now the problem that would be posed by our continued dependence on fossil fuels. We do now know, however, that the problems posed by the use of fossil fuels are probably not solvable while those based on nuclear fuels are. So in concluding this section on nuclear power, I also recommend "going for it" along with the caution to be described in the next paragraph.

This caution concerns the issue of "waste heat" referred to above in my discussion of solar-derived power. We humans use a little over 16 terawatts of power at any moment. If hoped-for developments of Class IV and beyond nuclear reactors become so successful that we find ourselves with an essentially unlimited supply of the heavy element fuels they require, then the amount of energy produced by those methods could soar to levels well beyond our current total global use of power. Since nuclear energy does not come from the sun, however, we will have to be mindful of the added amount of so called "waste energy" these non-solar sources provide (as is explained by Anil Anathaswamy in his article entitled "How Clean is Green" in

New Scientist, Jan., 2012, page35-38). If the total amount of energy used by man then increased, for example, by one hundred fold to about 1600 terawatts in say two centuries from now mainly by the adoption of nuclear technologies, the additional heat delivered to the Earth by those nuclear reactors would then increase the average temperature of the Earth by about 1.6 degrees Fahrenheit. This would constitute a dangerous level of additional temperature increase, approximately equal to that we have already experienced in the previous two centuries. Therefore, while we should replace our existing fossil-fuel-based power plants with nuclear plants in the short term — in order to meet our immediate needs — we should not become overly dependent on them in the longer term. For our longer term needs, we will increasingly have to depend on solar-based energy sources for which no waste energy penalty has to be paid. Thus, any technological advances we make now in the area of sunlight-derived power will serve us very well in both the short and long runs.

Energy efficiency, change, and (yes, oh my God, maybe even) sacrifice: Of all of the means we have at our disposal in order to address AGW, the "lowest hanging fruit" (that is, what we know for sure that we can do starting right now, if not before) is to increase our efficiency of energy use. Since it might be too scary to say here that we need to lower our "standard of living" in order to decrease our energy use, let's instead just say that we need to make changes in our "mode of living", while noting that our "quality of life" need not be diminished. In order to better understand this point, it is helpful to reflect on what has happened in our country during the last half of the twentieth century.

The industrial revolution gave the world the potential for unprecedented prosperity. The U.S. in particular had all the ingredients necessary to take full advantage of these changes. With our abundance of natural resources, a strong work ethic within our growing population, a free enterprise system, and a stable government, the U.S. became an industrial powerhouse that proved to be a deciding factor in World War II. After that war, our factories switched from making weapons of war to the manufacture of consumer goods. Within a few years in the 1950s, however, it became clear that our ability to produce these goods far exceeded the demand for them.

One of the ways this problem was then "solved" was by enormous advertising campaigns that induced people to buy things that they did not need and by increasing the availability of loans and credit to potential customers. As a result, the standard of living most Americans aspired to went far beyond that which would have otherwise been considered to be satisfactory. At the same time, it is not at all clear that these enormous material improvements between the 1950s and the 2000s have led to corresponding increases in our personal levels of contentment or "quality of living." My parents' generation clearly seemed to enjoy their lives in the '50s just as much as mine has enjoyed the most recent decade, fifty years later. Therefore, a redefinition of the standard of living we aspire to might not really be so painful for most of us and would definitely benefit the quality of living for future generations. The example of Warren Buffet comes to mind. While he is one of the wealthiest men in the world, he has lived simply and in the same modest home in Omaha, Nebraska since 1958 and appears to have enjoyed his life to the fullest. Note also that thanks to the advances in electronic and computing technologies made over the last half century, we now have access to countless sources of low-cost and low-impact entertainments and professional tools not available to previous generations.

In order to achieve our goals, it might also be necessary for many of us to transfer a substantial amount of our personal wealth to the collective wealth of our planet. This thought is not motivated by any preference for socialism, but also makes excellent business sense within the tenets of a free-market system that places value on its long-term as well as short-term interests. After all, Earth will not be worth much in the currency of human beings if it is allowed to be continuously degraded by AGW. We can prevent its future degradation only by an unprecedented level of collective action.

Which fossil fuels should we use during the transition period? We obviously cannot get to a completely fossil-fuel-free era overnight. That transition will take some time — perhaps the rest of this century. Nevertheless, it is exceedingly important that we use our existing supplies of fossil fuels wisely during this period. The challenge posed here is not that we will run out of fossil fuels. It is instead that if we are careless in our choice of the fuels we use, we will unnecessarily use too much of it.

The remainder of this section contains a lot of details and numbers. The math required to follow the conversation involves only simple addition and subtraction, however, and the answer to the question posed above is very clearly revealed in those numbers. The references for the numbers to be used here can be found in an article written in November, 2011, by Raymond Pierrehumbert, Professor of Geophysical Sciences at the University of Chicago. This article entitled "Keystone XL: Game Over?" can be found on the Web at www.realclimate.org. It deals specifically with the subject under consideration here and provides estimates of the available reserves of all forms of fossil fuels. Pierrehumbert is also a co-author of the 2011 National Research Council report entitled "Climate Stabilization Targets: Emissions, Concentrations, and Impacts over Decades to

Millennia". That document can also be found on the Web at www. nap.cdu/catalog.php?record_id=12877. While the latter document is exceedingly detailed and comprehensive, Pierrehumbert provides the following brief summary of its contents regarding CO_2 emissions and climate in his article cited above.

- The peak warming we get will be linearly proportional to the cumulative total amount of carbon dioxide (CO_2) emitted.

- It does not matter much how rapidly the CO_2 is emitted.

- The warming you get when you stop emitting CO_2 is what you are stuck with for the next thousand years.

- The climate recovers only slightly over the next ten thousand years.

- Using the mid-range of IPCC estimates for climate sensitivity, industrial era total CO_2 emissions of about 3,700 gigatons (equivalent to carbon emissions of 1,000 gigatons) will increase temperature by about 2.0 degrees Centigrade above the pre-industrial level.

In the ensuing discussion, it will be convenient to use units of Centigrade (C) for temperature change (a temperature change of 1.0 C is equal to a change of 1.8 degrees Fahrenheit) and gigatons (Gt) for expressing quantities of carbon (a gigaton is equal to one billion metric tons).

At the beginning of this discussion, it is useful to define a specific goal for minimizing future increases in temperature. Many scientists have set this maximum allowable temperature increase at about 2.0 C above the pre-industrial temperature. During the previous interglacial period, called the Eemian, that existed from about 130,000 to 114,000 years ago, the peak temperature is thought to have been about 2 to 3 C higher than our pre-industrial temperature. While physical conditions during the Eemian were similar to those of the Holocene (our present interglacial period), sea levels were then about 5 meters (16 feet) higher than today. Temperatures of 3 C higher than those of our pre-industrial era have probably not occurred in about 3 million years — just before the glacial/interglacial cycles began. At that time, sea levels were about 25 meters higher than today. In view of our planet's history, it appears that a goal of limiting temperature increase to 2.0 C might be too modest for comfort and I suspect that this goal is commonly used simply because it might be achievable. That is, this goal appears to be one born out of the realities of our present condition and not one that might have been selected a few decades ago. With that in mind, let's proceed with this discussion.

We also must recognize that the average temperature of the Earth has already increased about 0.8 C since the beginning of the Industrial Age — and we know that due to the thermal inertia of the oceans, we will experience another one to two tenths of a degree increase within the next couple decades from our present level of atmospheric CO_2. Thus, it appears that we have already caused an increase in the Earth's temperature of about 1.0 C. In addition, you will recall from our discussions of the Ice Core Record in Chapter 4 that the temperature sensitivity to CO_2 increases has two components, one is thought to be about 3.0 C for CO_2 doubling in the shorter term when fast feedbacks are operative, and another is about 6.5 C for the longer term when both fast and slow feedbacks become operative. Therefore, from the CO_2 levels currently in our atmosphere, we might expect close to another degree Centigrade increase in temperature within the next couple centuries due to the onset of the slow feedback effects associated with changes in our planet's sheet ice. Thus, the CO_2 we have already added to our atmosphere might be sufficient to get us most of the way to that defined limit of 2.0 C of warming.

Next, let's move on to consider our two very best and cleanest types of fossil fuels, natural gas and oil. It has been estimated (see Pierrehumbert's article referred to above), that we have about 100 Gt of natural gas and about 140 Gt of oil readily available for extraction and use throughout the world. Since we seem to be continuously finding more reserves of gas and oil each year, these numbers could easily increase by up to 50% within the next few decades. The sum of these two forms of fossil fuels thereby comes to about 240 Gt if we don't look too hard for more and up to about 360 Gt if we do.

Next, we should recognize that the world is very likely to use almost all of the gas and oil that it manages to find in the next century. This is because of the high energy content of gas and oil, their ease of extraction and distribution, and their cleanliness (with respect to not causing other environmental problems) relative to the other forms of fossil fuels. If we don't use the oil available in Saudi Arabia, for example, we must recognize that someone else will. And remember that AGW is a global and not a regional problem. As Benjamin Franklin said upon signing the Declaration of Independence, "We must hang together, gentlemen...else, we shall most assuredly hang separately."

Next, under the reasonable assumption that we will use our reserves of gas and oil, we should consider what impact that use will have on our atmosphere. We know that by burning about 500 Gt of fossil fuels over the Industrial Age so far, we have increased the level of atmospheric CO_2 from 280 to 393 ppm. In units of the atmospheric carbon content, that corresponds to a change from 550 to 780 Gt, an increase of about 230 Gt or 40%. These numbers suggest and other measurements have shown that about half of the carbon emitted by the combustion of fossil fuels stays in the atmosphere over the century it is emitted. Therefore, if we burn all of our readily available gas and oil during the present century, as I expect we will, we will be adding about 120 Gt (240 Gt x 0.5) of additional carbon to the atmosphere, resulting in a total atmospheric carbon load of 900 Gt (780 + 120). This amount of carbon corresponds to a CO_2 level of 458 ppm which is 45 ppm higher than today and 178 ppm higher than in 1850 for a total increase of 64% over the Industrial Age.

Alternatively, if we do manage to find 50% more gas and oil, we will then be emitting about 360 Gt of additional carbon from these two source in the next century, resulting in 180 Gt carbon added to the atmosphere, creating a total carbon atmospheric content of 960 Gt and a CO_2 level of about 490 ppm for a total increase of 75% over the Industrial Age.

At this point, let's reflect on what conditions were like when the Earth last had more than 450 ppm CO_2 in its atmosphere. That would be about 35 million years ago when the Earth had cooled just enough to begin forming sheet ice on the continent of Antarctica (see the Hansen *et al.* (2008) reference indicated in Suggested Reading for Chapter 3). The Earth was then still in an ice-free "water world" state with sea levels about 70 meters higher than today. Therefore, if we use only our currently available sources of gas and oil, we will still be headed in that direction — and it does appear that we will be doing just that during the current century. The only natural factor we will then have going for us will first be the temperature delaying effects of our oceans and then that of the Greenland and Antarctic ice sheets. Due to their great masses, it will take a couple decades to overcome the thermal inertia of the oceans and perhaps a couple centuries to overcome that of the ice sheets. Whether or not we would be able to reduce the heating effect of CO_2 during those precious few initial decades — perhaps by removing some CO_2 from the atmosphere and/ or by the implementation of other schemes for cooling — is presently unknown (again see Hansen *et al.* (2008) for thorough discussions of this point). Hansen *et al.* argue very convincingly that we must reduce levels of atmospheric CO_2 down to about 350 ppm during the current century in order to arrest our planet's drift towards its ice-free, "water world" condition. Considering that we already have 393 ppm CO_2 in our atmosphere and are heading towards at least 458 ppm, the task of returning to 350 ppm CO_2 will require nothing less than a Herculean effort.

Let's now move on to consider coal, the other widely used form of fossil fuel. The world is estimated to have about 650 Gt of coal in reserve and more of it is being discovered every year. So yes, coal really is the 800 gigaton gorilla in the fossil fuel room! Unless the technique of Carbon Capture and Sequestration (CCS) is made to be technically and financially feasible, however, it appears that the days of coal-fired power plants are numbered. In terms of its impact on land and its various emissions into the atmosphere, coal is also a relatively "dirty" source of power that provides much less energy per carbon atom emitted than do gas and oil. Even if some coal-fired power plants do manage to successfully develop CCS systems, the additional costs associated with CCS are likely to make those power plants financially unattractive relative to our other choices. For supplying baseline levels of power, for example, there will be even less doubt remaining that the development of Class III and Class IV nuclear power plants is a better way to go. Therefore, it appears that coal's future will be to provide much of our power, but only in the very immediate term — while we cut its use back very forcefully in the next decade and completely shut it down within a couple decades. Remember, that we will most likely be using all of our gas and oil. Therefore, any additional use of coal will simply add to the already excessive levels of atmospheric CO_2 we expect to get from gas and oil use.

Moving on to the other non-traditional sources, let's consider a distinctly lower grade of fossil fuel known as "tar sands". In the province of Alberta, Canada, alone, it is estimated that there are about 230 Gt of carbon in place within their tar sands. As described almost daily of late in our media, there is great interest in as well as great resistance to the construction of a long and large pipeline called the Keystone XL Pipeline that is proposed to connect the Alberta tar sands to the American refineries on the Gulf coast. From that point the refined products of the tar sands could be transported throughout the world. In view of the numbers I have described above, however, the obvious rhetorical question must be asked: why develop such a vast new source of carbon-based fuel when the world does not need it and, in fact, will be further damaged by its use? In addition, the other attendant environmental downsides associated with the production of oil from tar sands makes it at least as environmentally dirty as coal. Having spent four years of my life at the University of Alberta in Edmonton, I grew to love that province and its people. Nevertheless, concerning the further development of their tar sands — on which they have labored so diligently for half a century — I must say with all due respect and regret, "please stop, right now!" And concerning the timely question of whether or not the Keystone XL Pipeline should be built: "of course not!" Who needs it — other than the peddlers of the harmful and addictive substance this pipeline will spew onto the markets of the world? And sure the world needs more "jobs, jobs, jobs", but they should be directed towards saving, not destroying, human friendly conditions on our planet.

Lastly, just a few words about the abundant "shale oil" deposits found all over the world. By a very rough estimate, there appears to be just about as much of it on our planet as there is coal. Because of its very low quality and the environmental hazards it poses, its use for fuel has been limited. Given the numbers shown above, that is a good thing — shale should all be left in the ground, along with coal (unless used with CCS) and the tar sands.

So what do all the numbers provided above tell us? They tell us that during our transition from our technologies of the past — that were based on the energy derived from fossil fuels — to environmentally sustainable technologies, we should judiciously use only our remaining supplies of gas and oil and discontinue our use of all other forms of fossil fuels as quickly as possible. Ironically, the major reason for this is that there are (hopefully) limited supplies of gas and oil so that we will run out of them by the end of this century — while the supplies of the other forms of fossil fuels are essentially endless. Thus, we stand a chance of first arresting and then reversing the relentless advance of AGW only if we restrict our future use of fossil fuels to that of gas and oil.

Note that the numbers I provided above indicate that the total carbon we expect to have emitted over the industrial age will be about 740 Gt if we use only known sources of gas and oil and about 860 Gt if we find 50% more gas and oil. In addition, we will surely be using a significant amount of coal during the next decade or two. It should also be noted that we have not taken into account here the additional effects of the emissions of the stored terrestrial carbon that is likely to occur as the planet gets warmer (the Earth's vast regions of permafrost and ocean bottoms are known to hold great quantities of methane and carbon dioxide). The only reason we have not included these factors here is that we don't know enough about those processes yet to assign emission numbers to them. Thus, it appears that we will be getting uncomfortably close to or even exceeding the total carbon emissions level of 1000 Gt (indicated in the last bullet above) that is expected to result in a temperature increase of 2.0 C over the Industrial Age.

In concluding this section, the message from space-ship Earth clearly seems to be: "Houston, we have a problem". The next major phase of atmospheric research will surely focus on how we can avoid and/or survive the calamity we seem to be heading toward. Hopefully, we will be able to minimize the magnitude of that enormous problem by making judicious choices of the fossil fuels we use during our present period of transition to a fossil-fuel-free era.

Chapter summary: It is clear that a continuation of life as we have come to know it on this planet will require that we do all of the things described in this chapter while we simultaneously elevate our commitment to research and development within the basic and applied physical sciences. While the scientific understanding of how we got into our present state of peril is clear, the roadmap for how we are going to get out of it does not yet exist. We must hope that improved and perhaps even new technologies will be discovered soon while we also personally do our best to reduce carbon dioxide emissions and the relentless advance of AGW. If we are to stand any chance in this battle against AGW, it is absolutely essential that the most powerful country on the planet change its role from being the greatest part of the problem to becoming a major participant in its solution. In the past, the U.S. has demonstrated that it had the wisdom, courage and strength to rise to demanding but necessary challenges. Let's hope that we can do this at least one more time and thereby achieve our own "finest hour" within the history of civilization.

CHAPTER 10

The Best and Obvious Energy Policy

While our government currently passes out substantial amounts of money to various forms of both alternate and conventional sources of energy, we are not developing a sensible and sustainable national energy policy largely because the best and most obvious of policies is not being given a chance. So what is this best and most obvious energy policy?

First, it is not the Cap and Trade policy that you have probably heard a great deal about. I believe that the Cap and Trade approach is doomed from the start. This is partially because it is so susceptible to corporate lobbying, resulting in exemptions and opt-outs along with fraudulent offsets and carbon-credits that it is unlikely to provide credible and reliable approaches to climate change mitigation. Consider for example, would the EPA be able to correctly determine if planting a 600 tree forest in Nebraska constitutes an appropriate long-term carbon offset for a given emission exemption requested in Detroit? Would it then be able to verify that this forest is being appropriately developed and maintained in subsequent years? Also, wouldn't the monitoring of this long-term arrangement be quite expensive? Multiply these questions concerning just one project many thousands of times and I believe one quickly loses all faith in such a system.

Simply put, the best and obvious energy plan is the following: we should let the free market do its thing and charge the full costs of all energy sources. That's it. Sounds simple, right? And it's also the American way, right? So where's the hang up? The hang up is that the full cost of all energy schemes must include the costs associated with waste disposal. Well, of course, you might say — after all, waste disposal is a well known and large component of nuclear energy costs, right? So again, where is the hang up?

The hang up is that the users of fossil fuels have enjoyed free use of the Earth's atmosphere as a garbage dump for the disposal of carbon dioxide for more than a century and they do not want to start paying for that service now. The only defense they have for their position is to deny the now exceedingly well established science behind man-caused global warming. Unfortunately, the professional scientific organizations of the U.S.A have clearly stated that the costs of removing or adjustment to the extra carbon dioxide we continue to dump into our atmosphere every day will be enormous. Therefore, we must immediately create a level playing field for all means of energy production by charging all energy providers, including those in the fossil fuel businesses, the full costs of their product's use. Is this not, indeed, an obvious and fair way to proceed? Note that the only basis for claiming that this plan is not fair and appropriate is to join the choir of deniers of AGW who say that our increased levels of carbon dioxide do not pose a threat to the world's environment.

An optional, but also important aspect of the plan being described here is that 100% of the proceeds thereby collected from the carbon fees should be returned directly to the public. This is appropriate because our atmosphere is a classic example of a "common" property. It is analogous to a shared field of grass, for example, by which any individual rancher is bound to gain by adding more of his own cattle to it, even though all of the ranches will suffer from the overgrazing that might then result. Thus, the proceeds of the carbon fee should be distributed to members of the public on a per capita basis. This distribution could be easily done on a monthly basis via our existing IRS system. Hansen (2009) has provided the following example of how this system might affect an average American family of four. After a few years of gradual increase in this carbon fee (let's say when it has risen to about $115 per ton of carbon dioxide), the proceeds that would be returned to a typical family of four might then amount to about $750 per month or $9,000 per year ($3,000 per adult per year and half that to each of their dependents). That family could use those extra funds to purchase the then more expensive fossil fuels, if they wished (the price of gasoline would then have risen by about $1 per gallon), or they could try harder than they previously had to meet their energy needs in other ways and thereby pocket a portion of their dividend. Those alternative ways would be made increasingly attractive to them by a free market system in which the suppliers of gas, oil and coal would no longer be given free use of our atmosphere for waste disposal. In this way, public habits would be quickly transformed.

It should be noted here that the previous sentence also tells us why we have not yet been offered this plan by our elected officials. The producers of fossil fuels are terrified by the spread of this idea and, to a great extent, they control Congress. A particularly horrifying thought of the fossil fuel industries is that the value of the enormous fossil fuel reserves they presently hold and those they continue to look for would be substantially reduced by the imposition of a progressively increasing carbon fee. From the environmental standpoint, however, that is a good and necessary thing — most of our vast reserves of fossil fuels must stay in the ground.

With our present system, our government distributes money and grants to various alternate energy schemes and then wishes them good luck in their ensuing competition against the untaxed suppliers of gas, oil, and coal who, by the way, have received far greater levels of assistance from our government. As expected, most of the alternate energy schemes then fail after their government support is pulled. With a waste disposal fee added to all fossil fuel use, the cost of that use would be progressively increased each year, as it should be, and then the free market system alone would allow the best of our energy entrepreneurs to rise to the top and we would be well on our way to developing self-sustaining new energy industries. It is certainly true that there would be losers as well as winners in making these changes — just as there always have been. After the Model T Ford, for example, the blacksmiths of America did not do so well. On the other hand, one of our country's greatest industries was born. Is our country now suddenly going to become a net loser in order to sustain one of its out-dated industries? Hopefully, the industrial forces of America will not remain so stagnant.

Another reason why the fossil fuel industries hate this plan is that it is simple enough to be easily understood by everyone. With it, the only game to be played out in Washington DC would be that of annually adjusting the carbon fee. Presently, we have an impossibly complex system in which only the exceedingly dedicated and exceedingly wealthy can play. While I would like to think that our elected officials would welcome the simpler system being proposed here, I also suspect that far too many of them have grown to enjoy the personal power they presently wield in their dealings with well-healed lobbyists from both the fossil fuel and alternate energy industries.

The plan proposed here could also cause our warming problem to be addressed globally. This would be accomplished via the imposition of "carbon tariffs " on goods and services for which a carbon fee was not paid in the country of origin. In this way, other countries would undoubtedly begin to charge similar carbon fees internally so that their own governments would collect those fees instead of ours.

Note also that our government has been clearly and repeatedly informed of this plan — even though our elected officials have not generally shared it with the public. For example, a formal presentation of this plan, called Carbon Fee with 100% Dividend, was provided on February 25, 2009, before the House Ways and Means Committee by Dr. James Hansen, a leading climate change scientist. The full text of his presentation can be seen on the Web at: www.columbia. edu/~jeh1/mailings/2009/20090226_WaysAndMeans.pdf.

So why is this plan not embraced by more of our Congressmen? If you put that question to any of them, the answer you are likely to receive is simply this: "this plan has no chance of being politically acceptable in the U.S. Congress", period. That's it. The logic behind this anemic response is that big changes in existing business-as-usual practices are not likely to see the light of day in Washington because firmly entrenched existing businesses control Congress. This is a self-fulfilling prophecy that is not being sufficiently challenged today.

A good example of this is currently being played out in my home state of Montana. During this 2012 election year, a great deal of money is pouring into Montana from the outside and from Texas, in particular, in order to ensure that Montana's Senators do not develop too strong a conscience concerning the environmental downsides of ongoing fossil fuel developments in my state and our neighbor to the north, Alberta, Canada. So far, even our Democratic Senators and our Democratic Governor have been conducting themselves in a manner totally acceptable to the fossil fuel industries. In fact, they enthusiastically compete with our Republican Representative (and now senatorial candidate) in providing "leadership" for the forceful development of Montana's vast reserves of coal and Alberta's vast reserves of tar sands (yes, with "friends" like these, it is not at all surprising that appropriate leadership, so far, in the fight against AGW has been limited largely to professional scientists).

Therefore, until there is no longer any merit to self-fulfilling prophecies such as the operational one in Washington DC described above, significant changes will not occur. Instead the U.S. government is more likely to continue throwing money here and there at green energy suppliers in order to give the public the impression that they are doing something about the climate change problem – while simultaneously helping the fossil fuel industries find and develop still more sources of gas, oil, coal, and even dirtier and less energy-inefficient forms of fossil fuels such as the tar sands of Alberta. A new term has been coined for this behavior now so commonly displayed by our Democrats at all levels. It is called "greenwashing". Yes, it is true – on one side of our political aisle, we presently have the "deniers" of AGW and on the other we have "greenwashers". In one sense, the latter group sometimes does more harm than the former because they give the American public the mistaken impression that the AGW problem is being adequately addressed – while it definitely is not and our window of opportunity is quickly closing.

In summary, Will Roger's legendary observation, "we have the best Congress money can buy" is particularly sobering today because of the unprecedented and truly alarming dimensions of the climate change problem. Therefore, I hope that I have adequately explained here what obstacles need to be overcome in order to adopt the best and most obvious national energy policy. Getting there is a matter of finding the political will and leadership equal to the task. If children and the yet-to-be-born could vote, they would surely go for the Carbon Fee and 100% Dividend Plan. No other plan will do the job for them.

Definition of Terms and Acronyms

anthropogenic global warming (AGW): AGW is the warming of Earth due to the activities of man. Most of this is thought to be due to the combustion of fossil fuels and the resulting emissions of CO_2 into the atmosphere. Other contributions include the emissions of other greenhouse gases and the reduction of vegetation that removes CO_2 from the atmosphere.

albedo: This is the fraction of incoming solar radiation that is reflected by Earth's surfaces back into the universe. Snow-covered surfaces and clouds contribute major portions of this effect. For example, the solar radiation that strikes the snow-covered polar regions is largely reflected, making the albedo of such regions very high, about 80 percent. The albedo of water surfaces, on the other hand, is very low, less that 10 percent. The average albedo of Earth is thought to be about 30 percent. A decrease in the average albedo of Earth caused, for example, by the melting of the Arctic Ocean's ice cap would lead to an increase in Earth's average temperature.

chlorofluorocarbons (CFCs): Also known by their trade name, the

Freons, these compounds were produced for a variety of commercial and industrial purposes. They are no longer made in large quantities because they were shown to cause the destruction of our protective layer of ozone in Earth's stratosphere. They also act as greenhouse gases, and because of their exceedingly long atmospheric lifetimes, will remain in the atmosphere for several centuries to come.

greenhouse gases (GHGs): Greenhouse gases are components of the atmosphere that absorb infrared radiation emitted from Earth, thereby inhibiting Earth's means of cooling itself. As a result of this greenhouse effect, the temperature of Earth must increase so that the amount of infrared energy it emits into the universe is equal to the amount of energy it absorbs from the sun. Without the naturally occurring greenhouse gases, the average temperature of Earth would be about 60 °F colder than it is, much too cold to support many existing forms of life.

electromagnetic radiation (EMR): EMR is a form of energy emitted by all heated objects. The most familiar example of EMR is ordinary visible light, so named because our eyes can detect it. Visible light is emitted by the surface of the sun and any other object that is heated to a temperature of about 10,000 °F. Another example of EMR is infrared radiation, which is emitted by moderately warm objects such as a stove, your hand, or the Earth. The emission of EMR is how heated objects attempt to cool themselves.

forcing agent: This refers to a specific change in the existing conditions on Earth that will force its average temperature to either increase or decrease. For example, increased levels of carbon dioxide in the atmosphere will cause positive forcing by increasing the absorption of the infrared radiation coming from the Earth. Increased cloud formation can cause both negative forcing (by increasing the reflection of incoming solar radiation) and positive forcing (by absorbing the infrared radiation emitted by the Earth.

global dimming: This is a term used to describe a cooling effect caused by the reflection of incoming solar radiation by particulate matter (often referred to as aerosols) suspended in the atmosphere. Major sources of particulates include volcanoes, coal-fired power plants, and vehicles.

hydrofluorocarbons (HFCs): These man-made compounds have replaced the CFCs in many commercial and industrial applications because they do not destroy stratospheric ozone. They are greenhouse gases, however, and are presently accumulating in the atmosphere.

hydroxyl radical: This is a simple molecule (OH) consisting of one oxygen and one hydrogen atom that initiates the breakdown and removal of hydrocarbons in the troposphere. It is formed in the lower atmosphere by the interaction of the small amount of UV light that manages to penetrate through the stratosphere with tropospheric ozone and water vapor. Once made, hydroxyl radicals will quickly react with any molecule having a carbon-hydrogen bond (note that the CFCs do not have such bonds). The steady-state concentration of OH in the troposphere is exceedingly low (only about 0.01 pptr), and thereby, it provides a spectacular example of how even an ultra-trace component of the atmosphere can provide a vital atmospheric function; in this case, that of keeping it clean. It literally is the "janitor of the atmosphere."

ice ages: My use of this term here refers to what would be more rigorously called the "glacial periods" that have occurred during the last major ice age that we have been in for the last 2.5 million years. The warm periods between these glacial periods are rigorously called the "interglacial periods" of the last major ice age. Geologists tell us that there have been several major ice ages with the first occurring about 2.5 billion years ago.

infrared radiation (IR): IR radiation has wavelengths (1 to 50 microns) that are longer than those of visible light, and is emitted by all objects having moderate temperatures, (such as the Earth, itself). The emission of IR radiation is how the Earth attempts to cool itself. Infrared radiation is readily absorbed, however, by all molecules in the atmosphere that have 3 or more atoms (the greenhouse gases). Therefore, in order to emit enough IR radiation into the universe so as to maintain its total energy balance (solar energy "in" must equal IR energy "out"), the average temperature of the Earth must increase.

Intergovernmental Panel on Climate Change (IPCC): This is a scientific organization sponsored by the United Nations. It consists of the leading scientists in the world whose research areas involve climate change. They meet regularly and provide updated reports every three years. The contents of their last report, provided in 2007, are often referred to here.

lifetimes: This is a term I have used here to describe how long a given compound persists in the atmosphere. Stated more rigorously, this actually refers to the "half-life" of a given compound; that is, the time required for the concentration of that compound to be reduced to one-half its present value if further emissions of that compound were stopped.

microns: This is the unit of length used here to indicate the wavelength of the electromagnetic radiation. One micron is equal to one millionth of a meter, and one meter is just slightly longer (by 3.3 inches) than one yard. The width of a fine human hair is about 30 microns.

Milankovitch cycle: This term describes in detail how the Earth revolves around the sun. Over relatively long periods of time, small changes occur in the position and the tilt of the Earth relative to the sun. With respect to causing cold and warm periods of Earth, the most important factor is how much solar radiation strikes the Northern Hemisphere during its summer season. This is because the northern half of our planet has more land-versus-sea exposure than its southern half. If, due to the changes within Milankovitch cycles, the northern half gets less than its usual amount of irradiation during its summer season, a net increase in northern glaciations will occur, and this will cause the entire Earth to move toward a colder period. If the northern half gets more than its usual amount of summer irradiation, the entire Earth will move toward a warmer period.

ozone: Ozone (O_3) is the molecular form of oxygen having three atoms instead of two as in its normal form (O_2). Ozone is continuously formed by the interaction of ultraviolet radiation from the sun with O_2. The concentration of ozone is greatest in the middle of the stratosphere, but is also found at all other altitudes down to Earth's surface. The so-called "stratospheric ozone layer" serves the vital function of absorbing and thereby removing most of the ultraviolet light coming from the sun.

stratosphere: This is the portion of the atmosphere immediately above the troposphere. The air in this region is not vertically well mixed and, in fact, is stagnant due to its inverted temperature gradient (the temperature increases with increased altitude in the stratosphere). The stratosphere extends from the top of the troposphere up to an altitude of about thirty miles.

temperature scales: We will usually use the Fahrenheit ($^\circ$F) scale here because Americans are most familiar with it. Occasionally, the more scientifically preferred Celcius ($^\circ$C) scale will also be used. If a temperature *change* or *difference* is being referred to, that temperature change in units of $^\circ$C can be converted to units of $^\circ$F simply by multiplying by 1.8. If a *specific* temperature of some object is being indicated in units of $^\circ$C, that temperature can be converted to units of $^\circ$F by multiplying by 1.8 and then adding 32. Thus the freezing point of water (0.0 $^\circ$C) becomes 32.0 $^\circ$F and the boiling point of water (100 $^\circ$C) becomes 212 $^\circ$F. Note that the *change* in temperature when water is heated from its freezing point to its boiling point is +100 $^\circ$C or +180 $^\circ$F.

troposphere: This is the lower portion of Earth's atmosphere in which we live. It is well mixed from top to bottom and its height varies from about five miles in the polar regions to about twelve miles near the equator. The temperature at the top of the troposphere is very cold (approximately -70 $^\circ$F at mid-latitudes), thereby inhibiting the passage of water vapor and other compounds of moderate volatility into the stratosphere.

ultraviolet (UV) radiation: This is the short wavelength (0.1 to 0.4 microns) portion of the sun's incoming radiation. Unlike visible or infrared radiation, it can cause the rupture of the chemical bonds within most molecules.

visible radiation: Most of the electromagnetic radiation emitted by the sun is of this type, having wavelengths between 0.4 and 0.8 microns. As its name suggests, our eyes detect visible light.

Additional Reading

Chapter 1:

All aspects of Earth's atmosphere are well-described in the Web site of the University Corporation for Atmospheric Research (UCAR): *www.windows.ucar.edu.*

For a more rigorous description of the atmosphere: Peter V. Hobbs, *Introduction to Atmospheric Chemistry*, Cambridge University Press, Cambridge, MA, 2000.

For the basic science behind climate change: Robert Henson, *The Rough Guide to Climate Change*, Rough Guides Ltd., New York, 2006.

For the most authoritative and comprehensive information concerning climate change: The 2007 report of the United Nations' Intergovernmental Panel on Climate Change (the report of its Working Group I entitled "The Physical Science Basis" is particularly informative) at Web site: *www.ipcc.ch/ipccreports/ar4-wg1.htm.*

For a tour of the world, observing firsthand effects of climate change: Elizabeth Kolbert, *Field Notes from a Catastrophe*, Bloomsbury U.S.A, New York, 2006.

For an insightful overview of the Earth's attempt to respond to the effects of man: James Lovelock, *The Revenge of Gaia*, Basic Books, New York, 2006.

Chapter 2:

On what we can learn from ice core samples: Richard B. Alley, *The Two-Mile Time Machine: Ice Core, Abrupt Climate Change, and Our Future*, Princeton University Press, Princeton, NJ, 2000.

Chapter 3:

This very thorough account of what we have learned from ice and ocean bottom core samples is a "must read" for those with a scientific background: James Hansen *et al.*, Target Atmospheric CO_2: Where should Humanity Aim?, *The Open Atmospheric Science Journal*, vol. 2 (2008), pages 217-231.

A more public-friendly book by James Hansen providing similar scientific information as referenced above along with stories of Hansen's efforts to call the public's attention to AGW: James Hansen, *Storms of My Grandchildren*, Bloomsbury U.S.A, NY, 2009.

A history of the science of global warming: Spencer R. Weart, *The Discovery of Global Warming*, Harvard University Press, Cambridge, MA, 2003.

An earlier alarm for impending global warming: Stephan H. Schnieder, *Global Warming*, Sierra Club Books, San Francisco, 1989.

Chapter 4:

Documentation of the general agreement that exists among practicing scientists on the issue of AGW: Naomi Oreskes, "The Scientific Consensus on Climate Change," *Science*, vol. 306 (2004).

The complete U.S. Senate Minority Report, Dec. 11, 2008, is provided at the Web site *http://epw.senate.gov/public/index.*

cfm?FuseAction=Files.View&FileStore_id=83947f5d-d84a-4a84-ad5d-6e2d71db52d9.

Chapter 5:

An account of the CFC/stratospheric ozone controversy as well as other pollution problems: Mark Zachary Jacobson, *Atmospheric Pollution: History, Science, and Regulation*, 2nd Ed, Cambridge University Press, Cambridge, MA, 2002.

An account of the CFC/ozone controversy during its early stages: Lydia Dotto and Harold Schiff, *The Ozone War*, Doubleday & Company, Garden City, NY, 1978.

Chapter 6:

For more thoughts of Carl Sagan: Tom Head, editor, *Conversations with Carl Sagan*, University Press of Mississippi, Jackson, MS, 2006.

Chapter 7:

For a history of Winston Churchill during the period of 1932–1940: William Manchester, *The Last Lion*, Little, Brown and Company, Boston, MA, 1988.

Chapter 8:

For a description of the lay public's common reaction to AGW: Steve W. Running, *The Five Stages of Climate Grief*, can be found at the Web site, *ntsg.umt.edu/files/5StagesClimateGrief.htm.*

For an argument of why the wealthy must step up to the plate: Greg Palast, *How the Rich are Destroying the Earth*, Chelsea Green, White River Junction, VT, 2007.

Why it might already be too late to successfully address AGW:

James Lovelock, *Vanishing Face of Gaia*, Penguin Press, London, GB, 2009.

An appeal from a global warming activist: Laurie David, *Stop Global Warming: The Solution is You*, 2nd Edition, Fulcrum, Golden, CO, 2008.

Chapter 9:

For a forecast of the future effects of climate change based on the most recent scientific literature: Mark Lynas, *Six Degrees: Our Future on a Hotter Planet*. National Geographic, Washington, DC, 2008.

All aspects of global warming, including its likely consequences, are described in this well-illustrated guide to the findings of the IPCC: Michael E. Mann and Lee R. Kump, *Dire Predictions*, DK Publishing, New York, 2008.

For a concise summary of AGW solutions: Peter Barnes, *Climate Solutions: What Works, What Doesn't, and Why*, Chelsea Green, White River Junction, VT, 2008.

With the assistance of a team of scientists and engineers, former Vice President Al Gore describes technological approaches to solving AGW in *Our Choice: A Plan to Solve the Climate Crisis*, Rodale, 2009.

What individuals can do to prevent global warming: Diane G. McDilda, *365 Ways to Live Green*, Adams Media, Avon, MA, 2008.

What you can do to stop climate change or live through it: David De Rothchild, *Global Warming Survival Handbook*, Melcher Media, New York, 2007.

And finally, for those who care, this book provides an account of what might happen to the Earth if humans do not survive: Alan Weisman, *The World Without Us*, St. Martins Press, New York, 2007.

Chapter 10:

For a thorough discussion of the Carbon Fee and 100% Dividend policy, see Hansen's book referred to under Chapter 3. The version he presented to Congress in 2009 can be seen on the Web at: www. columbia.edu/~jeh1/mailings/2009/20090226_WaysAndMeans.pdf.

CPSIA information can be obtained at www.ICGtesting.com
Printed in the USA
LVOW12s2144121114

413436LV00001B/113/P